T0211677

SpringerBriefs in Molecular Science

Chemistry of Foods

Series Editor

Salvatore Parisi, Lourdes Matha Institute of Hotel Management and Catering Technology, Thiruvananthapuram, Kerala, India

The series Springer Briefs in Molecular Science: Chemistry of Foods presents compact topical volumes in the area of food chemistry. The series has a clear focus on the chemistry and chemical aspects of foods, topics such as the physics or biology of foods are not part of its scope. The Briefs volumes in the series aim at presenting chemical background information or an introduction and clear-cut overview on the chemistry related to specific topics in this area. Typical topics thus include:

– Compound classes in foods—their chemistry and properties with respect to the foods (e.g. sugars, proteins, fats, minerals, …)
– Contaminants and additives in foods—their chemistry and chemical transformations
– Chemical analysis and monitoring of foods
– Chemical transformations in foods, evolution and alterations of chemicals in foods, interactions between food and its packaging materials, chemical aspects of the food production processes
– Chemistry and the food industry—from safety protocols to modern food production

The treated subjects will particularly appeal to professionals and researchers concerned with food chemistry. Many volume topics address professionals and current problems in the food industry, but will also be interesting for readers generally concerned with the chemistry of foods. With the unique format and character of SpringerBriefs (50 to 125 pages), the volumes are compact and easily digestible. Briefs allow authors to present their ideas and readers to absorb them with minimal time investment. Briefs will be published as part of Springer's eBook collection, with millions of users worldwide. In addition, Briefs will be available for individual print and electronic purchase. Briefs are characterized by fast, global electronic dissemination, standard publishing contracts, easy-to-use manuscript preparation and formatting guidelines, and expedited production schedules.

Both solicited and unsolicited manuscripts focusing on food chemistry are considered for publication in this series. Submitted manuscripts will be reviewed and decided by the series editor, Prof. Dr. Salvatore Parisi.

To submit a proposal or request further information, please contact Dr. Sofia Costa, Publishing Editor, via sofia.costa@springer.com or Prof. Dr. Salvatore Parisi, Book Series Editor, via drparisi@inwind.it or drsalparisi5@gmail.com

Suni Mary Varghese · Salvatore Parisi ·
Rajeev K. Singla · A. S. Anitha Begum

Trends in Food Chemistry, Nutrition and Technology in Indian Sub-Continent

 Springer

Suni Mary Varghese
Lourdes Matha Institute of Hotel
Management and Catering Technology
Thiruvananthapuram, Kerala, India

Rajeev K. Singla ⓘ
Institutes for Systems Genetics, Frontiers
Science Center for Disease-related
Molecular Network
West China Hospital, Sichuan University
Chengdu, Sichuan, China

Salvatore Parisi
Lourdes Matha Institute of Hotel
Management and Catering Technology
Thiruvananthapuram, Kerala, India

A. S. Anitha Begum
Department of Home Science
Korambayil Ahammed Haji Memorial
Unity Women's College
Manjeri, Kerala, India

ISSN 2191-5407   ISSN 2191-5415   (electronic)
SpringerBriefs in Molecular Science
ISSN 2199-689X   ISSN 2199-7209   (electronic)
Chemistry of Foods
ISBN 978-3-031-06303-9   ISBN 978-3-031-06304-6   (eBook)
https://doi.org/10.1007/978-3-031-06304-6

This Springer imprint is published by the registered company Springer Nature Switzerland AG
The registered company address is: Gewerbestrasse 11, 6330 Cham, Switzerland

Contents

Chapter 1
Antioxidants and Nutritional Significance

Abbreviations

RCS   Reactive chlorine species
ROS   Reactive oxygen species
RNS   Reactive nitrogen species
RSS   Reactive sulphur species

1.1 Introduction to Antioxidants and Oxidative Stress

1.1.1 Foods Are a Complex Combination

The current situation of food and beverage industries and the market—from a general and 'global' viewpoint—is showing many interesting challenges for food technologists. A brief and absolutely not exhaustive list of arguments could contain many topics. Anyway, several points of discussion should be considered (Haddad and Parisi 2020; Parisi et al. 2021a, b, c):

(a)   The food or beverage product is the result of a transformation, except for a few situations. In other terms, we are speaking of transformed foods and/or beverages. Probably, different versions and sub-types of the same product (or the 'idea' of a general product) can be easily found depending on the geopolitical and traditional/historical area of certain regions and nations. Some works available in the recent literature have already reported examples of historical and technological contamination such as certain typical foods common to Italian or Arab cuisine. For example, the interesting example of Italian '*panissa*' (bread with slices), normally found in the Savona area, has been supposed to be the

S. M. Varghese et al., *Trends in Food Chemistry, Nutrition and Technology in Indian Sub-Continent*, Chemistry of Foods, https://doi.org/10.1007/978-3-031-06304-6_1

counterpart of another food specialty: the Sicilian 'pane e panelle' as the result of the presence of Genoese merchants in the Palermo area in the Middle Age (Barone and Pellerito 2020). Another interesting argument may be the discussion concerning Sicilian '*arancini*' of Catania, reproduced as form and purpose in Jordan when speaking of the traditional dairy *jameed* balls (Haddad et al. 2021a)

(b) Modern food and beverage products are protected, presented, and transported inside a container or composition of inedible, in general (plastic films, glass, metallic materials, etc.). This statement should highlight the role of 'invisible' and 'accessory' food packaging materials on the one side. On the other side, similar packaging/food integrations—the composed food products—do not include or mean only advantages. In fact, synergically joint containers (or associations of packaging materials) with contained foods or beverages could involve disadvantages intended as 'unexpected' food modifications. In general, we are speaking of the visible and detectable (by consumers) reaction of the food to new processes or treatments that would not have been possible in the past. An interesting and generally hidden aspect of the food technology—the needed presence and use of non-food components essential for food-grade machines such as food-grade lubricants should be highlighted and considered with attention

(c) Food safety-related scandals generally concern the detection—and health consequences—of non-chemical and biological risks: foreign bodies (detectable by means of metal detectors and X-ray inspection machines), and also chemical contaminants and biological toxins. These occurrences are the result of the sum of continuous and discontinuous sub-processes able to complete separately a food product along the whole food supply chain, from the primary production (agriculture, etc.) to the final distributor(s). As the number of separated operations grows, the complexity increases with the augment of possible disadvantages.

These reflections should be considered as a first attempt to discuss the role of chemical additives and ingredients, which are generally used in the food and beverage industry with the aim of obtaining more palatable and shelf-life stable products. In other terms, these—and other additives—are needed, but their presence is often not accepted by normal consumers, because 'the natural food is the best'. Such a definition of a food product is a real mistake at present: as recently affirmed, should a research chemist be obliged to show a food or beverage product without the presence of chemical compounds, He/She would be completely unable to satisfy this request. In other terms, there are no food or beverage products without chemical substances (Ballantini 2021).

What about chemical compounds (additives, surrogates, etc.) in the food industry? Their use is generally linked to easily detectable performances (by consumers): ameliorated (brilliant) colours, good smell, excellent texture, measurable and comparable chewingness… and other factors. Interestingly, these performances should be maintained within the so-called expiration date or use-by-date term, as a consequence of the First Parisi's Law of Food Degradation (Anonymous 2021; Parisi 2002; Srivastava 2019). However, the increasing worries of food consumers on the ground of human health have progressively generated other questions and requests. Consequently, the demand for 'natural' foods or 'clean labels' has increasingly augmented in recent years (Nieto et al. 2018; Parisi et al. 2021b), based on the possible use or enhancement of chemical additives with 'antioxidant power'. What could be told when speaking of this argument?

1.1.2   The Antioxidant Power

The use of chemical additives, especially from natural sources, has progressively increased in recent years because of three reasons, in particular (Barbieri et al. 2014; Barone et al. 2014; Nieto et al. 2018; Parisi and Dongo 2020; Parisi et al. 2020; Sahu et al. 2017; Singla 2020; Singla et al. 2021a, b; Sultana et al. 2021):

(1)   It has been reported that natural additives (extracts from vegetable organisms, etc.) can have a synergical action with preservation methods, which are common and historically accepted in the industry and by food consumers.
(2)   In addition, the 'natural' adjective is perceived as a synonym term for 'safe'.
(3)   Finally, several compounds have been specifically recommended in food applications because of their claimed antioxidant, anti-toxigenic, antibacterial, anti-diabetes, and antimutagenic features, and research still continues.

In other terms, these compounds have a large spectrum of health-positive properties. With reference to toxigenic, antibacterial, anti-diabetes, and antimutagenic features, adequate terminologies are available in the scientific literature. With concern to antioxidants, their action is generally named 'antioxidant power'. These words mean the power of these molecules against oxidant agents, from a chemical point of view. However, food technologists and biologists in the medical ambit above all have a profound interest in these compounds because 'antioxidant power' aims at contrasting biological damages, especially when speaking of attacks by reactive oxygen species (ROS). On the one side, antioxidant power is intended to limit oxidative reactions (with effects on texture, food performance, rancidity, and in general, durability) (Decker et al. 2005–2010); on the other side, antioxidants are used against biological damages, generally mentioned as 'oxidative stress' (Shahidi and Ambigaipalan 2015). In the ambit of this book, the second interpretation is considered. However, the 'natural' adjective should exclude the use of synthetic compounds, which are able to enhance antioxidant power in prepared foods and beverages. Consequently, our discussion should take into account a practical subdivision of antioxidant agents depending on their natural or artificial source.

1.2 Antioxidant Agents. Natural and Synthetic Compounds

Basically, there are two categories with concern to antioxidant agents of food interest (Flieger et al. 2021):

(1) Natural molecules: these compounds are further discriminated with reference to the endogenous or exogenous origin. In the first situation, the following sub-categories can be mentioned: enzymatic and non-enzymatic molecules.

(2) Synthetic molecules: these compounds are based on phenolic structures, polyphenols, etc., such as butylated hydroxyanisole, butylated hydroxytoluene, or *n*-propyl gallate. In this situation, health concerns have been considered so far because of possible contamination with chemical precursors, dangerous by-products, etc. On the other side, nano-antioxidants such as ascorbic acid in specific applications have been proposed. However, the synthetic origin and correlated health concerns have not favoured the use of similar compounds.

Consequently, the main research area concerning food and beverage applications includes natural-origin molecules, as shown in Fig. 1.1 (Flieger et al. 2021):

(a) Endogenous compounds: enzymatic and non-enzymatic species. The first group is further subdivided into: primary defence systems such as catalase or glutathione peroxidase, and secondary defence systems (i.e. glutathione reductase). On the other side, non-enzymatic species are discriminated on the basis

Antioxidant Agents. Natural Compounds

Enzymatic systems:
Catalase
Glutathione peroxidase
Glutathione reductase, ...

Non-enzymatic systems:
Lipoic acid
Uric acid
Albumin, ...

- Vitamins (ascorbic acid, retinol)
- Caroteinoids (lutein, lycopene, etc.)
- Trace metals (selenium, copper, zinc, manganese...)
- Polyphenols (flavanols, flavons, flavons, flavonoids, phenolics acids...)

Fig. 1.1 A simplified classification of natural antioxidants which may be used in the industry of foods and beverages

of low-molecular weight (lipophilic molecules, i.e. lipoic acid; or hydrophilic molecules, i.e. uric acid) or the ability of binding metals (i.e. albumin, ferritin, etc.)

(b)   Exogenous compounds: vitamins (ascorbic acid, retinol); carotenoids (lutein, lycopene, etc.); trace metals (selenium, copper, zinc, manganese…); and the large class of polyphenols (it is really larger than synthetic polyphenols because of the remarkable abundance in fruits and vegetables, and the diversification of chemical structures).

In relation to polyphenols, the diversification involves different subgroups such as flavanols, flavons, flavonoids such as anthocyanins, stilbenes, phenolic acids, etc. (Parisi and Dongo 2020).

1.3   Reactive Oxygen Species and Role of Antioxidants in Prevention of Diseases

The action of (preferably) natural antioxidants or synthetic compounds concerns essentially the delay of cell damages in the human body, contrasting 'oxidative stress'. This complex phenomenon can lead to the production of free radicals, which are highly unstable and can cause remarkable changes able to damage cells. Cellular and genetic changes may occur due to the highly reactive state of free radicals. These chemical species are substantially highly reactive compounds because of the existence of only one unpaired electron in the valence shell of the most interesting atom in the molecule. Because of this feature, these radicals can potentially act as oxidants (electron donors) or reducing species (electron acceptors). Anyway, these compounds are extremely dangerous because they are generally involved in metabolic processes such as reactions of the respiratory chain, prostaglandin synthesis, the cytochrome P-450 system, and other metabolic ambits (Flieger et al. 2021). In general, the medical discussion takes into account reactive oxygen species (ROS), i.e. hydroxyl, hydroperoxyl, alkyloxyl, and superoxide anion species. Actually, different non-radical species should be considered.

However, the existence of the following reactive compounds (Fig. 1.2) cannot be avoided (Flieger et al. 2021):

(a)   Reactive nitrogen species (RNS) such as radical nitric oxide (and other radicals and non-radical species);
(b)   Reactive chlorine species (RCS), such as radical chlorine or non-radical biatomic chlorine;
(c)   Reactive sulphur species (RSS) such as radical sulphur and non-radical hydrogen sulphide.

Because of the heterogeneity of complex oxidative-stress menaces (by the chemical and bio-chemical viewpoints), the discussion should take into account inflammations and cellular damages leading to cardiovascular disease, diabetes mellitus,

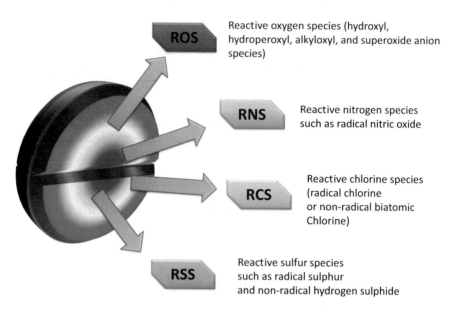

Fig. 1.2 Dangerous radical species

obesity, and Alzheimer's disease. During the normal process of ageing, antioxidant-rich diets can actively contribute to maintaining health and preventing diseases. The chemical action of dietary antioxidants is essentially a scavenging attitude against free radicals above all. As a result, intake of fresh vegetables and fruits, is extremely beneficial in preventing diseases caused by species such as ROS, RNS, RCS, and RSS (Haddad et al. 2021a, b).

The antioxidant potential of natural products, and even single chemicals, depends on many factors such as conditions of sample collections, as well as the extracts preparation method and the way of expressing results.

It has been recently reported that vegetable extracts can be not only used as sources of natural antioxidants 'as they are' (for direct use in food production) but also as sources for the production of nano-antioxidants. This line of research can be extremely promising because of environmental-friendly contents ('green synthesis' or 'sustainable chemistry') and relatively cheap solutions. In addition, these strategies could be used for therapeutic approaches in the medical sector with the possibility of direct supplementation by means of non-food applications (such as bandages). Consequently, the use of natural antioxidants should not be considered at present as a single approach for food production. The next chapters will discuss this and other topics (different active principles) in a more detailed way with concern to nutrition and diet-correlated ambits in the Indian subcontinent.

References

Anonymous (2021) Parisi's first law of food degradation valuable to establish adequate protocols concerning food durability. Inside Lab Manag 25(1):17

Ballantini V (2021) Professione Chimico 016—Il chimico in campo agroalimentare. https://youtu.be/aoMeIU3n5KE. Accessed 10 Dec 2021

Barbieri G, Barone C, Bhagat A, Caruso G, Conley ZR, Parisi S (2014) The influence of chemistry on new foods and traditional products. Springer International Publishing, Heidelberg, Cham. https://doi.org/10.1007/978-3-319-11358-6

Barone M, Pellerito A (2020) Palermo's street foods. The authentic arancina. In: Sicilian street foods and chemistry. Springer International Publishing, Cham, pp. 21–41. https://doi.org/10.1007/978-3-030-55736-2_2

Barone C, Bolzoni L, Caruso G, Montanari A, Parisi S, Steinka I (2014) Food packaging hygiene. Springer International Publishing, Cham. https://doi.org/10.1007/978-3-319-14827-4

Decker EA, Warner K, Richards MP, Shahidi F (2005) Measuring antioxidant effectiveness in food. J Agric Food Chem 53(10):4303–4310. https://doi.org/10.1021/jf058012x

Decker EA, Chen B, Panya A, Elias RJ (2010) Understanding antioxidant mechanisms in preventing oxidation in foods. In: Decker EA (ed) Oxidation in foods and beverages and antioxidant applications. Woodhead Publishing Series in Food science, technology and nutrition. Woodhead Publishing Limited, Sawston, pp 225–248. https://doi.org/10.1533/9780857090447.2.225

Flieger J, Flieger W, Baj J, Maciejewski R (2021) Antioxidants: classification, natural sources, activity/capacity measurements, and usefulness for the synthesis of nanoparticles. Mater 14(15):4135. https://doi.org/10.3390/ma14154135

Haddad MA, Parisi S (2020) The next big HITS. New Food Mag 23(2):4

Haddad MA, Yamani MI, Abu-Romman SM, Obeidat M (2021a) Chemical profiles of selected Jordanian foods. SpringerBriefs in Molecular science. Springer, Cham. https://doi.org/10.1007/978-3-030-79820-8

Haddad MA, Yamani MI, Da'san MMJ, Obeidat M, Abu-Romman SM, Parisi S (2021b) Food traceability in Jordan current perspectives. Springer International Publishing, Cham. https://doi.org/10.1007/978-3-030-66820-4

Nieto G, Ros G, Castillo J (2018) Antioxidant and antimicrobial properties of rosemary (Rosmarinus officinalis, L.): a review. Med 5, 3:98. https://doi.org/10.3390/medicines5030098

Parisi S (2002) Profili evolutivi dei contenuti batterici e chimico-fisici in prodotti lattiero-caseari. Ind Aliment 41(412):295–306

Parisi S, Dongo D (2020) Polifenoli e salute. I vegetali amici del sistema immunitario. Great Italian Food Trade. https://www.greatitalianfoodtrade.it/salute/polifenoli-e-salute-i-vegetali-amici-del-sistema-immunitario. Accessed 10 Dec 2021

Parisi S, Dongo D, Parisi C (2020) Resveratrolo, conoscenze attuali e prospettive. Great Italian Food Trade 27/10/2020. www.greatitalianfoodtrade.it/integratori/resveratrolo-conoscenze-attuali-e-prospettive. Accessed 10 Dec 2021

Parisi S, Liberatore G, Parisi C, Ballantini V (2021a) Il comparto agroalimentare. Chimica Industriale Essenziale, 26 maggio 2021. https://www.chimicaindustrialeessenziale.org/materiali-e-applicazioni/il-comparto-agrialimentare/. Accessed 10 Dec 2021

Parisi S, Parisi C, Liberatore G (2021b) Ingredienti naturali nell'industria. Il caso dei composti antiossidanti da fonti vegetali. Agro Food Ind Hi-tech 32, 3(Suppl):16–18

Parisi S, Varghese SM, Parisi C (2021c) The HITS strategy and chemometrics. Agro Food Ind Hi Tech 32(1):56–58

Sahu D, Sharma S, Singla RK, Panda AK (2017) Antioxidant activity and protective effect of suramin against oxidative stress in collagen induced arthritis. Eur J Pharm Sci 1(101):125–139. https://doi.org/10.1016/j.ejps.2017.02.013

Shahidi F, Ambigaipalan P (2015) Phenolics and polyphenolics in foods, beverages and spices: antioxidant activity and health effects—a review. J Funct Foods 18, B:820–897. https://doi.org/10.1016/j.jff.2015.06.018

Singla RK (2020) Secondary metabolites as treatment of choice for metabolic disorders and infec-
tious diseases and their metabolic profiling: Part 1. Curr Drug Metab 21(7):480–481. https://doi.
org/10.2174/138920022107200925101631

Singla RK, Ali M, Kamal MA, Dubey AK (2018) Isolation and characterization of nuciferoic acid, a
novel keto fatty acid with hyaluronidase inhibitory activity from Cocos nucifera Linn. Endocarp.
Curr Top Med Chem 18(27):2367–2378. https://doi.org/10.2174/1568026619666181224111319

Singla RK, Guimarães AG, Zengin G (2021a) Editorial: application of plant secondary metabolites to
pain neuromodulation. Front Pharmacol 11:623399. https://doi.org/10.3389/fphar.2020.623399

Singla RK, Gupta R, Joon S, Gupta AK, Shen B (2021b) Isolation, docking and in silico ADME-T
studies of acacianol: novel antibacterial isoflavone analogue isolated from Acacia leucophloea
Bark. Curr Drug Metab 22(11):893–904. https://doi.org/10.2174/1389200222666621005091417

Srivastava PK (2019) Status report on bee keeping and honey processing. MSME – Development
Institute, Ministry of Micro, Small & Medium Enterprises, Government of India 107, Industrial
Estate, Kalpi Road, Kanpur-208012. http://msmedikanpur.gov.in/cmdatahien/reports/diffIndus
tries/Status%20Report%20on%20Bee%20keeping%20&%20Honey%20Processing%202019-
2020.pdf. Accessed 10 Dec 2021

Sultana A, Singla RK, He X, Sun Y, Alam MS, Shen B (2021) Topical Capsaicin for the treatment
of neuropathic pain. Curr Drug Metab 22, 3:198–207. https://doi.org/10.2174/138920022199920
1116143701

Chapter 2
Phytochemicals

Abbreviations

–OH   Hydroxyl group
O_2 Molecular oxygen

2.1 A Brief Introduction to Phytochemical Substances

Foods and beverages are usually considered nutrient mixtures. In general, it can be inferred that edible materials have remarkable importance because of three essential functions for living organisms:

(a) Foods and beverages can sustain vital processes;;
(b) In addition, edible materials can sustain the growth
(c) And finally, these complex mixtures can supply energy (not only chemicals supposed to build/rebuild or repair tissues, organs, and so on).

The complexity of a food or beverage matrix relied on a roughly tripartite composition, with the exception of water as an essential dissolution medium (Chap. 3):

(1) Protein content,
(2) Fat (lipid) content,
(3) Carbohydrate content.

However, the remaining part may be essential for life. As a result, the rest of the food matrices could be roughly subdivided into two subcategories without a specific chemical characterization, but with only a discrimination factor: are these compounds essential for life, or not?

S. M. Varghese et al., *Trends in Food Chemistry, Nutrition and Technology in Indian Sub-Continent*, Chemistry of Foods,
https://doi.org/10.1007/978-3-031-06304-6_2

Should the answer be negative, we could consider many interesting food and beverage chemical profiles with some surprise, even if living organisms do not really need them to survive. This heterogeneous group contains many chemical species and subcategories, including 'phytochemicals'.

In general, these compounds are defined as chemicals produced by (and found in) vegetable organisms. From the above-mentioned viewpoint, these molecules are not essential nutrients. However, their importance should be considered from the viewpoint of nutritional significance and public health at least, even if there is limited evidence that certain molecules (or groups) have a specific action without synergy with other compounds found in 'healthy' foods such as fruits, pulses, whole grains, nuts, and certain vegetable products.

The real problem with phytochemicals is that their bioactivity is sure and well recognized; however, there is little information available with concern to the synergical action of these substances (and also the possible connection with other non-phytochemicals when speaking of beneficial effects on human health, at least). Why should these difficulties be discussed?

In fact, phytochemistry, the science of studying phytochemicals, has to rely on the knowledge and the continuous evaluation of six main pillars (Mendoza and Silva 2018):

(1) Chemical structures and configuration;
(2) Biosynthetic pathways and degradation mechanisms (in other words, metabolic processes);
(3) Natural distribution in vegetable organisms and fruits;
(4) Known biological functions and bioactive properties;
(5) And finally, methods of extraction and qualitative–quantitative analysis.

As a result, it should be admitted each new knowledge concerning phytochemicals is strongly dependent on the availability of plant materials (a good point, and the natural abundance is assured), the possibility of human and animal studies and researches (once more, this problem should be easily solvable), and... the availability of good extracts suitable for studies. However (Lu and Zhao 2017):

(a) Certain phytochemicals are photochemically unstable. In other words, these compounds may be easily photodegraded, in particular where the presence of photocatalyzing agents (also named sensitizers) is demonstrable. Easily visible and detectable food modifications—change of sensorial features, decrease of certain nutrients, production of toxic molecules—are only the most evident and 'obvious' result of photodegradation or photo-oxidation. In detail, three different mechanisms can be observed: direct absorption of light energy; electron transfer or proton transfer with the needed action of triplet-excited state sensitizers (this mechanism is named 'type I- photo-oxidation'); and/or photo-oxidation by means of the action of singlet oxygen produced by molecular oxygen (O_2) (this mechanism is named 'type II- photo-oxidation'). Anyways, phytochemical may be partially or totally destroyed... and this reflection should highlight the importance of good extraction and stabilization protocols in the laboratory and in the industry

(b)   Phytochemical sources—plant materials—may be stabilized by means of heat treatments (60 °C in ovens, etc.) until these tissues reach a constant weight. Subsequently, the following operations—extraction of phytochemicals (chromatographic systems, electrophoresis…); separation and isolation; identification of isolated molecules (ultraviolet spectroscopy, infrared spectroscopy, nuclear magnetic resonance spectroscopy, etc.); study of biosynthetic pathways for these phytochemicals; evaluation of quantities extractable from vegetable sources—have to be performed before considering animal or human experiments.

2.2 Classification of Phytochemicals. Two Possible Systems

The complexity of this matter needs some type of acceptable classification concerning phytochemicals before discussing their benefits. Two options can be offered here as a necessary premise (Fig. 2.1), while positive effects will be discussed in the next chapters (Higdon and Drake 2012, 2021; Higdon et al. 2021; Mendoza and Silva 2018; Merhan 2017):

(a)   First classification, based on three features only: biosynthetic pathways, structural features (concerning chemical structures), and solubility in water:

Phytochemicals. Two possible classifications...

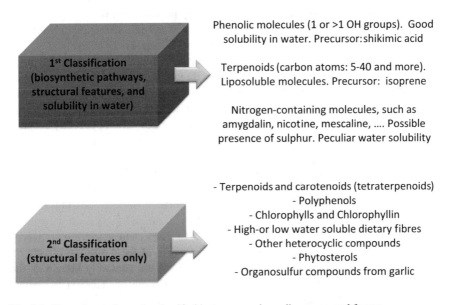

1st Classification (biosynthetic pathways, structural features, and solubility in water)

Phenolic molecules (1 or >1 OH groups). Good solubility in water. Precursor: shikimic acid

Terpenoids (carbon atoms: 5–40 and more). Liposoluble molecules. Precursor: isoprene

Nitrogen-containing molecules, such as amygdalin, nicotine, mescaline, …. Possible presence of sulphur. Peculiar water solubility

2nd Classification (structural features only)

- Terpenoids and carotenoids (tetraterpenoids)
- Polyphenols
- Chlorophylls and Chlorophyllin
- High- or low water soluble dietary fibres
- Other heterocyclic compounds
- Phytosterols
- Organosulfur compounds from garlic

Fig. 2.1 Phytochemicals can be classified in two ways depending on several factors

(a.1)   Phenolic molecules. One or more aromatic rings show one or more hydroxyl (-OH) groups. Their solubility in water is good. From the biochemical viewpoint, shikimic acid is considered the precursor.

(a.2)   Terpenoids (carbon atoms: 5–40 and more). These complex systems are derived from the precursor isoprene (which is repeated as the basic unit in their structure) and are liposoluble (i.e. isoprene, thymol, taxol, rubber, … depending on the increasing number of carbon atoms and the consequent number of isoprene units). From the biochemical viewpoint, two pathways are correlated with terpenoids: the glyceraldehyde phosphate–pyruvic acid pathway and the mevalonic acid pathway.

(a.3)   Nitrogen-containing molecules, such as cyanogenic glycosides (i.e. amygdalin), alkaloids (i.e. ephedrine, nicotine, mescaline, …). The presence of sulphur cannot be excluded. These structures present similarities with amino acids. In relation to bioactive alkaloids, their water solubility is a peculiar feature

(b)   Second classification, based mainly on the structural features of each class. Please consider that according to this classification, many species are defined also 'antioxidants' (Chap. 1):

(b.1)   With exclusive reference to carotenoids, these structures are organic pigments (ranging from yellow to red colours, and orange tints are normal). The most known examples are α- and β-carotene, lycopene (the red pigment of tomatoes), zeaxanthin (typically found in yellow sweetcorn and saffron), lutein (a xanthophyll representative, it is a yellow-coloured pigment commonly present on leaves of spinach and black pepper), etc. Because of their unsaturated and complex structures, they can act as excellent antioxidants (Chap. 1)

(b.2)   Polyphenols such as curcumin (typically found in turmeric), flavan-3-ols, anthocyanidins, flavanones, flavonols, flavones, isoflavones, stilbenes, and lignans. These compounds are generally ascribed to have strong or good antioxidant power (Chap. 1). Consequently, these substances—curcumin is a typical example—are apparently responsible for the most part of the beneficial effects on human health commonly associated with traditional medicines in Asia. In general, anti-inflammatory, anti-diabetes, anti-tumoural, and neuroprotective activity has been ascribed and often reported when speaking of polyphenols (Coniglio et al. 2018; Haddad et al. 2020a; Laganà et al. 2017, 2019, 2020; Singla et al. 2019). A peculiar mention should be made in relation to stilbenes, and resveratrol (3,4',5-trihydroxystilbene) in particular because of their positive effects on the cardiovascular system. These effects are generally known as the 'French Paradox' (Coniglio et al. 2018; Haddad et al. 2020a, b; Laganà et al. 2017, 2019, 2020; Parisi and Dongo 2020; Singla et al. 2019). In addition, a peculiar class of phenolic compounds—soy isoflavones—are claimed to act as 'phytoestrogens': this word means that these molecules show oestrogen-agonist and oestrogen-antagonist features at the same time. More research is needed in this ambit when speaking of anticancer properties, especially with reference to breast tumours

(b.3) Chlorophylls and chlorophyllin. The first molecules, called chlorophyll a and b, are natural liposoluble pigments, while chlorophyllin should be considered a semi-synthetic material obtained by mixing copper and sodium salts from chlorophyll. This mixture is water-soluble. With reference to possible health effects on human health, in particular concerning the possible action against some tumours, it has to be considered that these compounds may 'block' certain dangerous chemicals by means of chemical complexes. However, there is little information so far, especially in relation to the possible and concomitant complexation of nutrients (with consequent decrease in nutritional intakes)

(b.4) Dietary fibres, which can be further discriminated on the basis of their high solubility in water (i.e. psyllium, one of the most known components of anti-constipation mixtures), or low aqueous solubility (i.e. wheat bran, generally known as a popular breakfast product). Chemically, these fibres are complex carbohydrates and lignins. Human's digesting enzymes cannot demolish them: for this reason and their adsorbing properties against one of cholesterol forms and glucose, the supplementation of certain fibres can be useful when speaking of glycaemic control, contrasting actions against some tumours of the human digestive apparatus, and constipation (depending on their water solubility).

(b.5) Other heterocyclic compounds such as indole-3-carbinol and isothiocyanates, are generally correlated or derived from glucosinolates. These molecules, commonly found in *Brassica* vegetables, are known as possible agents able to indirectly influence the metabolism and removal of carcinogens. Antioxidant and anti-inflammatory properties are also ascribed to these heterocyclic structures

(b.6) Sterols of vegetable origin (phytosterols). These structures, similar to cholesterol, are known because of their capability to displace cholesterol from micelles, with consequent positive effects on the cardiovascular system. On the other hand, excessive supplementation of phytosterols could reduce carotenoids in human plasma, with possible negative effects

(b.7) Organosulfur compounds from garlic: these molecules derive from L-cysteine (sulfoxides such as alliin, γ-glutamyl-L-cysteine peptides, etc.). In this ambit, more research is still needed because these compounds are claimed to have antioxidant, antimicrobial, anti-inflammatory, and anti-tumoural activity. In addition, cardiovascular diseases could be contrasted with the supplementation of these molecules. However, clear evidences are not still reported so far.

This preliminary discussion is needed because of the importance of phytochemicals in the ambit of nutrition requirements, positive health effects, and diet counselling (Chaps. 7 and 8). There are many scientific papers and literature speaking and reporting positive effects (including ethnomedicinal, phytochemical, and ethnopharmacological significance) of these and other compounds of vegetable origin both in the Indian ambit only (Abat et al. 2017; Ananth et al. 2019; Babu et al. 2018; Cervera-Mata et al. 2021; Deepa et al. 2013; Kondhare and Lade 2017; Krishnaswamy and Raghuramulu 1998; Rao 2003; Sen et al. 2020; Sharma et al. 2019a, 2021) and on a global scale (Dillard and German 2000; Issaoui et al. 2020a, b; Johns 1996; Johns and Chapman 1995; Laganà et al. 2019, 2020; Lillehoj et al. 2018; Liu 2003; Ma and Zhang 2017; Mallik et al. 2020; O'Shea et al. 2012; Oz and Kafkas 2017; Prakash et al. 2012; Shakir and Rashid 2019; Sharma et al. 2019b). The interested reader is invited to consult also these references because the argument is extremely vast and variegated, and the next chapters could not be fully exhaustive in this broad ambit of research.

References

Abat JK, Kumar S, Mohanty A (2017) Ethnomedicinal, phytochemical and ethnopharmacological aspects of four medicinal plants of Malvaceae used in Indian traditional medicines: a review. Med 4(4):75. https://doi.org/10.3390/medicines4040075

Ananth DA, Deviram G, Mahalakshmi V, Sivasudha T, Tietel Z (2019) Phytochemical composition and antioxidant characteristics of traditional cold pressed seed oils in South India. Biocatal Agric Biotechnol 17:416–421. https://doi.org/10.1016/j.bcab.2018.12.018

Babu KN, Hemalatha R, Satyanarayana U, Shujauddin M, Himaja N, Bhaskarachary K, Kumar BD (2018) Phytochemicals, polyphenols, prebiotic effect of Ocimum sanctum, Zingiber officinale, Piper nigrum extracts. J Herb Med 13:42–51. https://doi.org/10.1016/j.hermed.2018.05.001

Cervera A, Sahu PK, Chakradhari S, Sahu YK, Patel KS, Singh S, Towett EK, Martín-Ramos P, Quesada-Granados JJ, Rufián JA (2021) Plant seeds as source of nutrients and phytochemicals for the Indian population. Int J Food Sci Technol 57(1):525–532. https://doi.org/10.1111/ijfs.15414

Coniglio MA, Laganà P, Faro G, Marranzano M (2018) Antimicrobial potential of Sicilian Honeys against Staphylococcus aureus and Pseudomonas aeruginosa. J AOAC Int 101(4):956–959. https://doi.org/10.5740/jaoacint.17-0450

Deepa G, Ayesha S, Nishtha K, Thankamani M (2013) Comparative evaluation of various total antioxidant capacity assays applied to phytochemical compounds of Indian culinary spices. Int Food Res J 20(4):1711–1716

Dillard CJ, German JB (2000) Phytochemicals: nutraceuticals and human health. J Sci Food Agric 80(12):1744–1756. https://doi.org/10.1002/1097-0010(20000915)80:12%3C1744::AID-JSFA725%3E3.0.CO;2-W

Haddad MA, Dmour H, Al- JFM, Obeidat M, Al- A, Al- AN, Al- MS, Shatnawi MA, Iommi C (2020a) Herbs and medicinal plants in Jordan. J AOAC Int 103(4):925–929. https://doi.org/10.1093/jaocint/qsz026

Haddad MA, El- J, Abu- S, Obeidat M, Iommi C, Jaradat DSM (2020b) Phenolics in Mediterranean and Middle East important fruits. J AOAC Int 103(4):930–934. https://doi.org/10.1093/jaocint/qsz027

Higdon J, Drake VJ (2012) Evidence-based approach to phytochemicals and other dietary factors. Thieme Publishers, New York

Higdon J, Drake VJ (2021) Chlorophyll and chlorophyllin. Linus Pauling Institute, Oregon State University, Corvallis. https://lpi.oregonstate.edu/mic/dietary-factors/phytochemicals/chlorophyll-chlorophyllin. Accessed 11 Dec 2021

Higdon J, Drake VJ, Delage B (2021) α-carotene, β-carotene, β-cryptoxanthin, lycopene, lutein, and zeaxanthin. Linus Pauling Institute, Oregon State University, Corvallis. https://lpi.oregonstate.edu/mic/dietary-factors/phytochemicals/carotenoids. Accessed 11 Dec 2021

Issaoui M, Delgado AM, Caruso G, Micali M, Barbera M, Atrous H, Ouslati A, Chammem N (2020a) Phenols, flavors, and the mediterranean diet. J AOAC Int 103(4):915–924. https://doi.org/10.1093/jaocint/qsz018

Issaoui M, Delgado AM, Iommi C, Chammem N (2020b) Polyphenols and the Mediterranean diet. Springer Nature Switzerland AG, Cham. https://doi.org/10.1007/978-3-030-41134-3

Johns T (1996) Phytochemicals as evolutionary mediators of human nutritional physiology. Int J Pharmacogn 34(5):327–334. https://doi.org/10.1076/phbi.34.5.327.13254

Johns T, Chapman L (1995) Phytochemicals ingested in traditional diets and medicines as modulators of energy metabolism. In: Arnason JT, Mata R, Romeo JT (eds) Phytochemistry of medicinal plants. Springer, Boston, pp 161–188

Kondhare D, Lade H (2017) Phytochemical profile, aldose reductase inhibitory, and antioxidant activities of Indian traditional medicinal Coccinia grandis (L.) fruit extract. 3 Biotech 7, 378:1–10. https://doi.org/10.1007/s13205-017-1013-1

Krishnaswamy K, Raghuramulu N (1998) Bioactive phytochemicals with emphasis on dietary practices. Indian J Med Res 108:167–181

Laganà P, Anastasi G, Marano F, Piccione S, Singla RK, Dubey AK, Delia S, Coniglio MA, Facciolà A, Di Pietro A, Haddad MA, Al- M, Caruso G (2019) Phenolic substances in foods: Health effects as anti-inflammatory and antimicrobial agents. J AOAC Int 102(5):1378–1387. https://doi.org/10.1093/jaoac/102.5.1378

Laganà P, Avventuroso E, Romano G, Gioffré ME, Patanè P, Parisi S, Moscato U, Delia S (2017) Chemistry and hygiene of food additives. Springer International Publishing, Cham. https://doi.org/10.1007/978-3-319-57042-6

Lagana P, Coniglio MA, Fiorino M, Delgado AM, Chammen N, Issaoui M, Gambuzza ME, Iommi C, Soraci L, Haddad MA, Delia S (2020) Phenolic substances in foods and anticarcinogenic properties: a public health perspective. J AOAC Int 103(4):935–939. https://doi.org/10.1093/jaocint/qsz028

Lillehoj H, Liu Y, Calsamiglia S, Fernandez- ME, Chi F, Cravens RL, Oh S, Gay CG (2018) Phytochemicals as antibiotic alternatives to promote growth and enhance host health. Vet Res 49:76. https://doi.org/10.1186/s13567-018-0562-6

Liu RH (2003) Health benefits of fruit and vegetables are from additive and synergistic combinations of phytochemicals. Am J Clin Nutr 78(3):517S-520S. https://doi.org/10.1093/ajcn/78.3.517S

Lu B, Zhao Y (2017) Photooxidation of phytochemicals in food and control: a review. Ann New York Acad Sci 1398(1):72–82. https://doi.org/10.1111/nyas.13377

Ma ZF, Zhang H (2017) Phytochemical constituents, health benefits, and industrial applications of grape seeds: a mini-review. Antioxid 6(3):71. https://doi.org/10.3390/antiox6030071

Mallik S, Sharangi AB, Sarkar T (2020) Phytochemicals of coriander, cumin, fenugreek, fennel and black cumin: a preliminary study. Natl Acad Sci Lett 43(5):477–480. https://doi.org/10.1007/s40009-020-00884-5

Mendoza N, Silva EME (2018) Introduction to phytochemicals: secondary metabolites from plants with active principles for pharmacological importance. In: Asao T, Asaduzzaman (eds) Phytochemicals: source of antioxidants and role in disease prevention. IntechOpen, London, pp. 25–48. https://doi.org/10.5772/intechopen.78226

Merhan O (2017) The biochemistry and antioxidant properties of carotenoids. In: Cvetkovic D, Nikolic G (eds) Caroteonids, pp 51–66. https://doi.org/10.5772/67592

O'Shea N, Arendt EK, Gallagher E (2012) Dietary fibre and phytochemical characteristics of fruit and vegetable by-products and their recent applications as novel ingredients in food products. Innov Food Sci Emerg Technol 16:1–10. https://doi.org/10.1016/j.ifset.2012.06.002

Oz AT, Kafkas E (2017) Phytochemicals in fruits and vegetables. In: Waisundara V (ed) Superfood and functional food. IntechOpen, London, pp 175–184. https://doi.org/10.5772/66987

Parisi S, Dongo D (2020) Polifenoli e salute. I vegetali amici del sistema immunitario. Great Italian Food Trade. https://www.greatitalianfoodtrade.it/salute/polifenoli-e-salute-i-vegetali-amici-del-sistema-immunitario. Accessed 10th Dec 2021

Prakash D, Gupta C, Sharma G (2012) Importance of phytochemicals in nutraceuticals. Jf Chin Med Res Develop 1(3):70–78

Rao BN (2003) Bioactive phytochemicals in Indian foods and their potential in health promotion and disease prevention. Asia Pac J Clin Nutr 12(1):9–22

Sen S, Chakraborty R, Kalita P (2020) Rice-not just a staple food: a comprehensive review on its phytochemicals and therapeutic potential. Trends Food Sci Technol 97:265–285. https://doi.org/10.1016/j.tifs.2020.01.022

Shakir BK, Rashid RMS (2019) Physiochemical and phytochemical profile of unripe black grape juice (verjuice). Ann Tropical Med Health 22, IV:S359. https://doi.org/10.36295/ASRO.2019.22126

Sharma RK, Micali M, Pellerito A, Santangelo A, Natalello S, Tulumello R, Singla RK (2019a) Studies on the determination of antioxidant activity and phenolic content of plant products in India (2000–2017). J AOAC Int 102(5):1407–1413. https://doi.org/10.5740/jaoacint.19-0136

Sharma DR, Kumar S, Kumar V, Thakur A (2019b) Comprehensive review on nutraceutical significance of phytochemicals as functional food ingredients for human health management. J Pharmacogn Phytochem 8(5):385–395. https://doi.org/10.22271/phyto.2019.v8.i5h.9589

Sharma RK, Micali M, Rana BK, Pellerito A, Singla RK (2021) Indian herbal extracts as antimicrobial agents. In: Indian herbal medicines. SpringerBriefs in Molecular Science. Springer Nature Switzerland AG, Cham, pp 31–40. https://doi.org/10.1007/978-3-030-80918-8_2

Singla RK, Dubey AK, Garg A, Sharma RK, Fiorino M, Ameen SM, Haddad MA, Al- M (2019) Natural polyphenols: chemical classification, definition of classes, subcategories, and structures. J AOAC Int 102(5):1397–1400. https://doi.org/10.1093/jaoac/102.5.1397

Chapter 3
Health Benefits of Non-nutrients. Dietary Fibre and Water

Abbreviations

NSP   Non-starch polysaccharide
H_2O Water

3.1   Dietary Fibres and Water. Are They Needed in Human Nutrition?

Nonessential nutrients can be synthesized by the human body. As a result, food is not the only possible origin. As briefly discussed in Chap. 2, foods and beverages are considered nutrient mixtures because of three essential functions for living organisms. These edible materials are needed when speaking of sustainability of vital processes and growth of humans and animals (with reference to the animal kingdom). Moreover, their importance is notable because of the need for bioavailable energy. In fact, living organisms must take energy from each possible source, and this energy has to be promptly available. As a result, foods and beverages are needed not only for building/repairing vital organs, tissues, and so on but also as energetic sources.

In addition, and according to current technological knowledge, the composition of foods and beverages is generally identified with a tripartite structure composed of lipids (fat matters), protein, and carbohydrates. However, this description is not correct because one peculiar and absolutely needed component is absent. This component—water—is essential because of two reasons:

(1)   All living microorganisms need life for survival.
(2)   Secondly, all foods and beverages are solid, semi-solid, colloidal, or liquid aqueous solutions. Consequently, and taking into account also dried/semi-dried foods, it has to be considered that water is needed and always present, even at

© The Author(s), under exclusive license to Springer Nature Switzerland AG 2022
S. M. Varghese et al., *Trends in Food Chemistry, Nutrition and Technology in Indian Sub-Continent*, Chemistry of Foods,
https://doi.org/10.1007/978-3-031-06304-6_3

very low concentrations (in lyophilized products, a minimal aqueous content is still present).

In other words, foods (and beverages above all) are water-dissolved solutions. This is a normal condition and also a pre-requisite for all edible products in human history, also meaning that dehydration treatments may be considered as one of the many tracts of anthropic activities and civilization in general. The simple production of more or less dehydrated cheeses (high-durability preserved milk) in many countries and areas, such as the well-known *jameed* balls in Jordan or certain high-dry matter Italian cheeses (Barone et al. 2014; Haddad and Parisi 2020; Haddad et al. 2020a,b, 2021a, b; Parisi 2006; Parisi et al. 2004), should easily demonstrate this fact. In addition, the addition of water to certain products has to be mentioned from two opposite viewpoints (Hading 1995; Khan et al. 1999; Pereira et al. 2006; Poonia et al. 2017):

(a) The regulatory perspective, with particular reference to food adulterations by means of the addition of liquid water to foods, especially in the milk and dairy sector.

(b) The technological perspective, with particular reference to the need for enhanced palatability for certain products by means of the addition of fluids (and also water).

From the economical viewpoint, water is also extremely convenient because of its very low price compared with other raw materials. The simple sentence *'Water is Life'* may have not only a purely biological meaning, but also an explicit meaning when speaking of economically convenient products and reliable savings concerning production costs for food business operators (even if water addition is clearly mentioned). In addition, water is needed in the food industry in relation to production purposes without a nutritional significance. As a result, water is needed… as a technological and economical factor, and its nutritional importance is naturally included (Delgado et al. 2017; Laganà et al. 2017; Mania et al. 2018; Sharma et al. 2019). However, water is also essential for life, and the same thing can be affirmed with reference to other non-nutrients of vegetable origin (phytochemicals) briefly discussed in Chap. 2. In the ambit of phytochemistry, the class of 'dietary fibres' has been also mentioned because of:

(1) Their role as health promoters in the human diet;
(2) Their contrasting importance against certain health disorders;
(3) Their high or low water solubility;
(4) And finally, their resistance—when observable—to human-produced digesting enzymes.

Each of these points has to be examined with connection to the role of water in the human diet and foods/beverages in general.

3.2 Definition and Functional Properties of Water as Nutrient

From the viewpoint of human nutrition, water (H_2O) has six important functions (Jéquier and Constant 2010):

(1) H_2O can be considered as a material needed for building human tissues and organs. This affirmation should be evaluated both on a macromolecular scale (with concern to tissues such as human skin) and a micromolecular scale, when speaking of single cells. In fact, the intercellular space is practically a liquid solution, and intercellular exchanges cannot be performed in other ways. In addition, water is contained in each cell.

(2) Secondly, the 'obvious' function of H_2O as a solvent medium should be clear enough. This function is observed regularly in natural foods. However, water can also be extremely important from the biochemical viewpoint because all reactions used for sustaining life need a reaction and solvent medium at the same time. As a result, H_2O is needed also as a partner (reactive) species for chemical reactions. In addition, pH conditions are extremely important when speaking of reaction yields, and water—an amphoteric molecule because it exhibits acidic and basic features at the same time—can play a critical role.

(3) Moreover, H_2O can act as an excellent transport medium (carrier) both for nutrient molecules and toxic catabolites (waste results from biochemical reactions) at the same time. This function, generally ignored, is absolutely critical because all vital organs and systems of the human body have to rely on the efficient transportation of nutrients such as carbohydrates, salts, etc., on the one side, and on the prompt elimination of toxic compounds and catabolites, on the other side. Consequently, ready nutrition must rely on water through the interstitial fluid (Grandjean et al. 2003), and the excretion of catabolites has to follow the same 'water ways' in the organism. Dehydration phenomena are extremely dangerous and potentially life-threatening.

(4) Another important function of water is thermoregulation (Montain et al. 1999). In fact, H_2O can contain a good energetic amount of heat. As a result, water can be used with the aim of reducing thermal modifications in the human body against adverse environmental (extreme cold or warm) conditions. The perspiration process (elimination of water solutions containing catabolites) from skins can be very useful if the human body is forced to reduce its thermal amount in a relatively small time temperature (Montain et al. 1999). When sweating is elicited, evaporation of water from the skin surface is a very efficient way to lose heat.

(5) H_2O can have a critical function in various environments of the human body as a lubricant agent. Actually, the role of lubricant has to be assured provided that water dissolves viscous compounds with the aim of creating a lubricating medium where requested.

(6) Finally, H_2O is needed as a simple and little molecule able to model and 'protect' the cellular structures against mechanical shocks. This important

function, often neglected, is extremely evident when speaking of pregnancies and possible damages to the foetus. In this situation, an aqueous 'wall' can protect the new organism from shocks.

These roles can explain well the importance of H_2O in the human body, and also the need for an efficient hydration over a 24-h period, so that losses and gains are approximately equal. For this reason, the recommended dietary intake of liquid water (excluding foods and other beverages) should be 1.5 L per day, excluding sedentary people (2.0–2.9 L). The amount of aqueous content in the human body is approximately 60% and it decreases continually over time, from the infant to adult age. In the last situation, it can be important to note that extracellular water is approximately 33% of the total amount. As a result, approximately 450 ml/day of H_2O are lost by evaporation from the skin, while normal respiration and solid excretion are responsible for an additional 300 and 200 ml/day of H_2O elimination, respectively (Kleiner 1999; Jéquier and Constant 2010).

3.3 Definition, Types, Structures, and Functions of Dietary Fibres as Nutrients

In general, dietary fibres are defined (DeVries 2003) as a fraction of (edible) vegetable organisms, which can be assimilated into a carbohydrate structure. This condition is needed but insufficient: the other necessary condition is that these carbohydrate matrices can resist the enzymatic digestion in the human small intestine with consequent fermentative reactions in the large intestine. In other words, dietary fibres cannot be absorbed through small intestine walls with the consequent destination to the large intestine and subsequent fermentation. Interestingly, laxative functions and other roles fibres can act as nutrient- and catabolite-absorbers (by means of complexation phenomena and other approaches) depending on the fermentation with consequent effects on intestinal pH and the production of useful by-products (from the physiological point: cholesterol and glucose control in the blood).

As a result, the composition of dietary fibre relies on the abundance of cellulose and hemicellulose (with associated fibrous aspect). The importance of dietary fibres can be discussed at least from the viewpoint of vegetable sources and/or with reference to their chemical and technological/functional characterization, as reported often in the scientific literature (Boers et al. 2017a, b; Boukid et al. 2019; Davis et al. 2018; Fahim et al. 2019; Jan et al. 2017; Kalala et al. 2018; Lila et al. 2017; Madane et al. 2020; Millar et al. 2019; Priya et al. 2019; Qi and Tester 2019; Sasaki et al. 2018; Sofi et al. 2017; Van Soest and Jones 2021; Venkidasamy et al. 2019; Vijayastelar et al. 2017). In detail, the complete list should be more challenging, as briefly explained here by means of a block-like classification (Buttriss and Stokes 2008; Ha et al. 2000):

First block: non-starch polysaccharides (NSP) and digestion-resistant oligosaccharides;

Second block: 'analogous carbohydrates';
Third block: lignins, (when associated with NSP and lignin complexes).

The first block concerns water-insoluble cellulose, soluble and insoluble hemicellulose, and other non-digestible polysaccharides. According to the above-mentioned definition, these (generally) water-insoluble compounds—polyglucoses, polyfructoses, β-glucans, different heteropolymers, soluble mucilages, pectins (relatively water-soluble and gel-forming complexes), and soluble gums—are fermentable. The second block concerns water-soluble analogous carbohydrates produced by means of physical or chemical reactions. This step occurs normally during food processing operations, and it influences the digestion of starches. Substantially, the main and necessary feature of 'analogous carbohydrates' (polydextrose; methyl cellulose; dextrins from maize, potatoes, etc.; synthetic carbohydrates; digestion-resistant starches) is the resistance to digestion, followed by difficult fermentation. Finally, the third block (lignins, intrinsically linked with hemicelluloses in spite of their non-carbohydrate nature) includes a heterogeneous group of water-insoluble natural polymers: waxes, cutin, suberin (fatty acid derivatives with strong resistance to digestion: these substances remain relatively undegraded by the microbial population in the large intestine), saponins, tannins, etc.(Ha et al. 2000).

Anyways, and with specific relation to potential health effects on human health, it has to be considered that dietary fibres are reported to have beneficial effects when speaking of cardiovascular diseases, glycaemic control, and gastrointestinal functionality. Actually, there is need for further studies in this ambit, because the group of dietary fibres is extremely variegated and also correlated with a wide—too wide— range of vegetables, from whole grains to red fruits (Mann and Cummings 2009). It is essential to recognize that these effects are probably linked with solubility in water. In fact, and as briefly anticipated in Chap. 2, these complex compounds could be differentiated on the basis of their high solubility in water (e.g. psyllium, one of the most known components of anti-constipation mixtures), or low aqueous solubility (e.g. wheat bran). On the other side, it has been reported that the soluble/insoluble differentiation may not have a remarkable importance on the practical ground because all undigestible fibers have a variable degree of resistance to digestion. With reference to fermentation and claimed physiological effects, the above-mentioned discrimination has not been demonstrated to have important and conclusive influences (Buttriss and Stokes 2008). The lack of evidence is more important at present because of the wide range of food products with claimed properties justified by the presence of dietary fibres, and the problem of reliable food labelling is a current issue, nowadays. Another worry depends on the removal of fibres from a number of transformed food products, with associated effects on the human health and the possibility of digestion-related disorders, tumours, and other diseases (the opposite situation, abundance of added fibres, is reported to be correlated with the reduction of coronary heart diseases, obesity, and some tumours. However, more research is still needed.

Certainly, the use of dietary fibres is well-promoted in the food industry even if the possible health effects are not researched as the first objective. In fact, dietary fibres actively improve and modify the textural appearance of foods and other organoleptic features because of their strong gel-forming attitude with water and also anti-sticking and anti-clumping features. In addition, cooking yields and fat binding are reported to be enhanced if dietary fibres are added to the food mass mixture. For these reasons, the research should investigate the use and consequences of dietary fibres in the industry because of their wide applications.

References

Barone C, Bolzoni L, Caruso G, Montanari A, Parisi S, Steinka I (2014) Food packaging hygiene. Springer International Publishing, Cham. https://doi.org/10.1007/978-3-319-14827-4

Boers HM, MacAulay K, Murray P, Dobriyal R, Mela DJ, Spreeuwenberg MA (2017a) Efficacy of fibre additions to flatbread flour mixes for reducing post-meal glucose and insulin responses in healthy Indian subjects. Brit J Nutr 117(3):386–394. https://doi.org/10.1017/S00071145170 00277

Boers HM, van Dijk TH, Hiemstra H, Hoogenraad AR, Mela DJ, Peters HP, Vonk RJ, Priebe MG (2017b) Effect of fibre additions to flatbread flour mixes on glucose kinetics: a randomised controlled trial. Brit J Nutr 118(10):777–787. https://doi.org/10.1017/S0007114517002781

Boukid F, Zannini E, Carini E, Vittadini E (2019) Pulses for bread fortification: a necessity or a choice? Trends Food Sci Technol 88:416–428. https://doi.org/10.1016/j.tifs.2019.04.007

Buttriss JL, Stokes CS (2008) Dietary fibre and health: an overview. Nutr Bull 33(3):186–200. https://doi.org/10.1111/j.1467-3010.2008.00705.x

Davis KF, Chiarelli DD, Rulli MC, Chhatre A, Richter B, Singh D, DeFries R (2018) Alternative cereals can improve water use and nutrient supply in India. Sci Adv 4, 7:eaao1108. https://doi.org/10.1126/sciadv.aao1108

Delgado AM, Almeida MDV, Parisi S (2017) Chemistry of the mediterranean diet. Springer International Publishing, Cham. https://doi.org/10.1007/978-3-319-29370-7

DeVries JW (2003) On defining dietary fibre. Proc Nutr Soc 62(1):37–43. https://doi.org/10.1079/PNS2002234

Fahim JR, Attia EZ, Kamel MS (2019) The phenolic profile of pea (Pisum sativum): a phytochemical and pharmacological overview. Phytochem Rev 18:173–198. https://doi.org/10.1007/s11101-018-9586-9

Grandjean AC, Reimers CJ, Buyckx ME (2003) Hydration: issues for the 21st century. Nutr Rev 61(8):261–271. https://doi.org/10.1301/nr.2003.aug.261-271

Ha MA, Jarvis MC, Mann JI (2000) A definition for dietary fibre. Eur J Clin Nutr 54(12):861–864. https://doi.org/10.1038/sj.ejcn.1601109

Haddad MA, Dmour H, Al-Khazaleh JFM, Obeidat M, Al-Abbadi A, Al-Shadaideh AN, Al-mazra'awi MS, Shatnawi MA, Iommi C (2020a) Herbs and medicinal plants in Jordan. J AOAC Int 103(4):925–929. https://doi.org/10.1093/jaocint/qsz026

Haddad MA, El-Qudah J, Abu-Romman S, Obeidat M, Iommi C, Jaradat DSM (2020b) Phenolics in mediterranean and middle east important fruits. J AOAC Int 103(4):930–934. https://doi.org/10.1093/jaocint/qsz027

Haddad MA, Parisi S (2020) The next big HITS. New Food Mag 23(2):4

Haddad MA, Yamani MI, Abu-Romman SM, Obeidat M (2021a) Chemical profiles of selected Jordanian foods. Springer Briefs Mol Sci Springer, Cham. https://doi.org/10.1007/978-3-030-79820-8

Haddad MA, Yamani MI, Da'san MMJ, Obeidat M, Abu-Romman SM, Parisi (2021b) Food traceability in jordan current perspectives. Springer International Publishing, Cham. https://doi.org/10.1007/978-3-030-66820-4

Harding F (1995) Adulteration of milk. In: Milk quality, pp 60–74. Springer, Boston. https://doi.org/10.1007/978-1-4615-2195-2_5

Jan R, Saxena DC, Singh S (2017) Effect of extrusion variables on antioxidant activity, total phenolic content and dietary fibre content of gluten-free extrudate from germinated Chenopodium (Chenopodium album) flour. Int J Food Sci Technol 52(12):2623–2630. https://doi.org/10.1111/ijfs.13549

Jéquier E, Constant F (2010) Water as an essential nutrient: the physiological basis of hydration. Eur J Clin Nutr 64(2):115–123. https://doi.org/10.1038/ejcn.2009.111

Kalala G, Kambashi B, Everaert N, Beckers Y, Richel A, Pachikian B, Neyrinck AM, Delzenne NM, Bindelle J (2018) Characterization of fructans and dietary fibre profiles in raw and steamed vegetables. Int J Food Sci Nutr 69(6):682–689. https://doi.org/10.1080/09637486.2017.1412404

Khan M, Rajah KK, Haines M (1999) Quantitative techniques in the measurement of milk adulteration in Peshawar. Pakistan. Int J Dairy Technol 52(1):20–25. https://doi.org/10.1111/j.1471-0307.1999.tb01989.x

Kleiner SM (1999) Water: an essential but overlooked nutrient. J Am Diet Assoc 99(2):200–206. https://doi.org/10.1016/S0002-8223(99)00048-6

Laganà P, Avventuroso E, Romano G, Gioffré ME, Patanè P, Parisi S, Moscato U, Delia S (2017) chemistry and hygiene of food additives. Springer International Publishing, Cham. https://doi.org/10.1007/978-3-319-57042-6

Lila MA, Schneider M, Devlin A, Plundrich N, Laster S, Foegeding EA (2017) Polyphenol-enriched berry extracts naturally modulate reactive proteins in model foods. Food Funct 8(12):4760–4767. https://doi.org/10.1039/C7FO00883J

Madane P, Das AK, Nanda PK, Bandyopadhyay S, Jagtap P, Shewalkar A, Maity B (2020) Dragon fruit (Hylocereus undatus) peel as antioxidant dietary fibre on quality and lipid oxidation of chicken nuggets. J Food Sci Technol 57(4):1449–1461. https://doi.org/10.1007/s13197-019-04180-z

Mania I, Delgado AM, Barone C, Parisi S (2018) food additives in cheese substitutes. In: Traceability in the dairy industry in Europe, pp 109–117. Springer International Publishing, Cham. https://doi.org/10.1007/978-3-030-00446-0_6

Mann JI, Cummings JH (2009) Possible implications for health of the different definitions of dietary fibre. Nutr Metab Cardiovasc Dis 19(3):226–229. https://doi.org/10.1016/j.numecd.2009.02.002

Millar KA, Gallagher E, Burke R, McCarthy S, Barry-Ryan C (2019) Proximate composition and anti-nutritional factors of fava-bean (Vicia faba), green-pea and yellow-pea (Pisum sativum) flour. J Food Comp Anal 82:103233. https://doi.org/10.1016/j.jfca.2019.103233

Montain SJ, Latzka WA, Sawka MN (1999) Fluid replacement recommendations for training in hot weather. Mil Med 164(7):502–508. https://doi.org/10.1093/milmed/164.7.502

Parisi S (2006) Profili chimici delle caseine presamiche alimentari. Ind Aliment 45(457):377–383

Parisi S, Laganà P, Delia S (2004) Cubed mozzarella cheese in modified atmosphere packages: evolutive profiles of chemical and microbiological parameters during shelf life. In: Proceedings of the 3rd international symposium on food packaging, Barcelona, Spain, 17–19 November 2004, Programme & Abstracts Book. International Life Science Institute (ILSI) Europe, Brussels

Pereira LJ, de Wijk RA, Gavião MBD, van der Bilt A (2006) Effects of added fluids on the perception of solid food. Physiol Behav 88(4–5):538–544. https://doi.org/10.1016/j.physbeh.2006.05.005

Poonia A, Jha A, Sharma R, Singh HB, Rai AK, Sharma N (2017) Detection of adulteration in milk: a review. Int J Dairy Technol 70(1):23–42. https://doi.org/10.1111/1471-0307.12274

Prasad VSS, Hymavathi A, Babu VR, Longvah T (2018) Nutritional composition in relation to glycemic potential of popular Indian rice varieties. Food Chem 238:29–34. https://doi.org/10.1016/j.foodchem.2017.03.138

Priya TR, Nelson ARLE, Ravichandran K, Antony U (2019) Nutritional and functional properties of coloured rice varieties of South India: a review. J Ethn Foods 6:1–11. https://doi.org/10.1186/s42779-019-0017-3

Qi X, Tester RF (2019) Utilisation of dietary fibre (non-starch polysaccharide and resistant starch) molecules for diarrhoea therapy: a mini-review. Int J Biol Macromol 122:572–577. https://doi.org/10.1016/j.ijbiomac.2018.10.195

Sasaki D, Sasaki K, Ikuta N, Yasuda T, Fukuda I, Kondo A, Osawa R (2018) Low amounts of dietary fibre increase in vitro production of short-chain fatty acids without changing human colonic microbiota structure. Sci Rep 8:435. https://doi.org/10.1038/s41598-017-18877-8

Sharma SD, Bhagat AR, Parisi S (2019) Water, carbon, and phosphorus footprint concerns in the food industry. In: Raw material scarcity and overproduction in the food industry, pp 13–29. Springer International Publishing, Cham. https://doi.org/10.1007/978-3-030-14651-1_2

Sofi SA, Singh J, Rafiq RR (2017) Fortification of dietary fiber ingredients in meat application: a review. Int J Biochem Res Rev 19(2):1–14. https://doi.org/10.9734/IJBCRR/2017/36561

VanSoest PJ, Jones LHP (2021) Analysis and classification of dietary fibre. In: Brätter P, Schramel P (eds) Proceedings of the fifth international workshop "trace element analytical chemistry in medicine and biology", vol 5, Neuherberg, Federal Republic of Germany, April 1988, pp 351–370. De Gruyter, Berlin and New York. doi: https://doi.org/10.1515/9783112417201-038

Venkidasamy B, Selvaraj D, Nile AS, Ramalingam S, Kai G, Nile SH (2019) Indian pulses: a review on nutritional, functional and biochemical properties with future perspectives. Trends Food Sci Technol 88:228–242. https://doi.org/10.1016/j.tifs.2019.03.012

Vijayasteltar L, Jismy IJ, Joseph A, Maliakel B, Kuttan R, Krishnakumar IM (2017) Beyond the flavor: a green formulation of Ferula asafoetida oleo-gum-resin with fenugreek dietary fibre and its gut health potential. Toxicol Rep 4:382–390. https://doi.org/10.1016/j.toxrep.2017.06.012

Chapter 4
Food Additives

Abbreviation

Citric acid   E330

4.1 Food Additives. Why?

Food additives can be defined as chemical substances deliberately added to foods directly or indirectly in known and regulated quantities for the purposes of assisting in the processing of foods, preservation, and improving flavor, texture, and/or appearance of foods. Food additives have an important role in the production of processed products. Food additives are unavoidable in the complex and integrated society of urbanization to maintain distribution networks by adding preservatives. At present, there is a great demand for convenience and ready-to-serve food products so food additives are essential to prevent rancidity and durability performances of foods. Additives play a key role in maintaining food qualities and characteristics that consumers demand, keeping food safe, wholesome, and appealing from farm to fork.

This brief introduction should be discussed not only from the technological viewpoint, but also from the viewpoint of public health and food safety. On the one side, a large portion of foods and beverages could not be realized and placed on world markets without the use of food additives, in spite of 'obvious' worries by different consumers (Ballantini 2021; Laganà et al. 2017). On the other side, their role is to modify and preserve food and beverage quality and safety. This objective has to be obtained by means of the improvement of the 'designed' food or beverage product against several organic modifications (Laganà et al. 2017):

© The Author(s), under exclusive license to Springer Nature Switzerland AG 2022   25
S. M. Varghese et al., *Trends in Food Chemistry, Nutrition and Technology in Indian Sub-Continent*, Chemistry of Foods,
https://doi.org/10.1007/978-3-031-06304-6_4

(a)   Each type of natural reaction is favoured by thermal fluctuations over time, even if thermal preservations systems (0–4 °C-refrigeration or freezing methods) are used.
(b)   Many of the above-mentioned chemical reactions are performed or are in synergy with microbial activity. Consequently, the chemistry of foods and beverages is biochemistry, after all.
(c)   Finally, several of the above-mentioned reactions–oxidation, colorimetric modifications, textural changes—are easily observable by normal consumers. As a result, the use of several additives may not only preserve foods over time, but also preserve external appearance at the same time.

4.2   Food Additives. Why Not?

Substantially, food additives are a strong way to obtain safe foods, and marketable products, and also a powerful instrument when speaking of the fight against world scarcity of foods, provided their safety and food-grade status are demonstrated (Archer et al. 2008; Ayala-Zavala et al. 2011; Barbera 2020; Barone et al. 2014; Bowles and Juneja 1998; Christaki et al. 2013; Deshpande and Deshpande 2017; European Parliament and Council 2010; Fiorino et al. 2019; Issaoui et al. 2020; Kumar et al. 2000; Laganà et al. 2019; Mania et al. 2018; Mudgil et al. 2011; Murcia et al. 2004; Nigam and Luke 2016; Nithin et al. 2020; Palaniyappan et al. 2013; Rather et al. 2016; Saltmarsh 2000; Satyanarayana et al. 2004; Sharma 2018; Singh et al. 2003; Singla et al. 2019; Soni et al. 1997). On the other side, the 'excessive' use of these may be easily questioned and criticized in many ambits (Barlow 1990; Matta and Gjyli 2016; Metcalfe et al. 2011; Silva and Lidon 2016). As a simple example, the use of citric acid (E330), a simple acidity corrector, can be generally considered a fraud agent when speaking of adulteration of 'authentic' lemon juices (McDonald 2020). Anyways, one of the major concerns related to food additives is their possible effect on food safety. A recent example is the ban of titanium dioxide as a food additive in the European Union in early 2022 (Askew 2021). Consequently, a little discussion concerning the use and classification of food additives in the current food and beverage industry should be made now.

4.3   A Brief Classification of Additives for Food and Beverage Productions

A simplified discrimination of food additives may be offered depending on their intended/declared function. In this ambit, the following classification can be discussed (Laganà et al. 2017):

(a)   Food preservatives,
(b)   Sensorial enhancers,

(c)   Technological enhancers,
(d)   Sweetness enhancers.

The first group concerns the protection of the food/beverage products against microbial activity and oxidant agents. Antimicrobials can be natural and traditional (sodium chloride, acetic acid…) or synthetic (benzoic acid, natamycin, nitrites and nitrates, E code from 200 to 297) compounds (Laganà et al. 2017). In relation to antioxidants (already discussed in Chaps. 1 and 2), their additive code (E300 to E330) concerns the general protection of food texture and other sensorial features by means of the inhibition of oxidant reactions. These compounds (L-ascorbic acid, some gallates, butyl hydroxyanisole, the above-mentioned citric acid, etc.) are oxidized by the oxidant agent(s) instead of the food or beverage itself.

With reference to sensorial enhancers, on the other side, natural, natural-identical, or synthetic dyes can be used (E 100 to E 199) depending on the need for colorimetric enhancement of the food or beverage products (Laganà et al. 2017). In the same ambit, flavor agents (E 620–E 640) can enhance the odor features of the product. These compounds may be also considered part of technological enhancers (with stabilizers, thickeners, gelling agents, humectants, etc.) because one of the most evident results is the natural appearance of the food or beverage as perceived by the normal consumer. Finally, sweetness enhancers have to be considered. Actually, the concern is focused on synthetic sweeteners. On the other side, a public-health perspective could not consider the excessive use of natural sweeteners in the food diet because of the occurrence of diseases correlated with the consumption of carbohydrates and also fat matter (Laganà et al. 2017). For this reason, and not only with concern to the problem of synthetic additives as opposed to natural ones, the matter of these compounds will be discussed in the subsequent chapters from a specific safety and prevention viewpoint, while the chemical discussion should take into account other parameters.

References

Archer DB, Connerton IF, MacKenzie DA (2008) Filamentous fungi for production of food additives and processing aids. In: Stahl U, Donalies UE, Nevoigt E (eds) Food biotechnology. Advances in biochemical engineering/biotechnology, vol 111. Springer, Berlin, Heidelberg. https://doi.org/10.1007/10_2007_094

Askew K (2021) https://www.foodnavigator.com/Article/2021/10/11/EU-E171-ban-set-for-2022-The-safety-of-our-food-is-not-negotiable. William Reed Business Media Ltd. https://www.foodnavigator.com/Article/2021/10/11/EU-E171-ban-set-for-2022-The-safety-of-our-food-is-not-negotiable.

Ayala-Zavala J, Vega-Vega V, Rosas-Domínguez C, Palafox-Carlos H, Villa-Rodriguez JA, Siddiqui MW, Dávila-Aviña JE, González-Aguilar GA (2011) Agro-industrial potential of exotic fruit byproducts as a source of food additives. Food Res Int 44(7):1866–1874. https://doi.org/10.1016/j.foodres.2011.02.021

Ballantini V (2021) Professione chimico 016-Il chimico in campo agroalimentare. https://youtu.be/aoMelU3n5KE.

Barbera M (2020) Reuse of food waste and wastewater as a source of polyphenolic compounds to use as food additives. J AOAC Int 103(4):906–914. https://doi.org/10.1093/jaocint/qsz025

Barlow SM (1990) Toxicological aspects of antioxidants used as food additives. In: Hudson JF (ed) Food antioxidants. Springer, Dordrecht, pp 253–307

Barone C, Bolzoni L, Caruso G, Montanari A, Parisi S, Steinka I (2014) Food packaging hygiene. Springer International Publishing, Cham. https://doi.org/10.1007/978-3-319-14827-4

Bowles BL, Juneja VK (1998) Inhibition of foodborne bacterial pathogens by naturally occurring food additives. J Food Saf 18(2):101–112. https://doi.org/10.1111/j.1745-4565.1998.tb00206.x

Christaki E, Bonos E, Giannenas I, Karatzia MA, Florou-Paneri P (2013) Stevia rebaudiana as a novel source of food additives. J Food Nutr Res 52(4):195–202

Deshpande A, Deshpande B (2017) Food additives and preservation: A review. Indian J Sci Res 13:219–225

European parliament and council (2010) commission regulation (EU) No 257/2010 of 25 March 2010 setting up a program for the re-evaluation of approved food additives in accordance with regulation (EC) No 1333/2008 of the European parliament and of the council on food additives. Off J Eur Union L 80:19–27

Fiorino M, Barone C, Barone M, Mason M, Bhagat A (2019) Quality systems in the food industry. Springer International Publishing, Cham. https://doi.org/10.1007/978-3-030-22553-7

Issaoui M, Delgado AM, Iommi C, Chammem N (2020) Polyphenols and the mediterranean diet. Springer Nature Switzerland AG, Cham. https://doi.org/10.1007/978-3-030-41134-3

Kumar JK, Sharma AK, Kulkarni PR (2000) Effect of preservation techniques and food additives on staphylococcal thermonuclease. Food Nahr 44(4):272–275. https://doi.org/10.1002/1521-3803(20000701)44:4%3C272::AID-FOOD272%3E3.0.CO;2-Q

Laganà P, Avventuroso E, Romano G, Gioffré ME, Patanè P, Parisi S, Moscato U, Delia S (2017) Chemistry and hygiene of food additives. Springer International Publishing, Cham. https://doi.org/10.1007/978-3-319-57042-6

Laganà P, Campanella G, Patanè P, Cava MA, Parisi S, Gambuzza ME, Delia S, Coniglio MA (2019) Chemistry and Hygiene of Food Gases. Springer International Publishing, Cham. https://doi.org/10.1007/978-3-030-35228-8

Mania I, Delgado AM, Barone C, Parisi S (2018) Traceability in the dairy industry in Europe, Springer international publishing, Cham, pp. 109–117. https://doi.org/10.1007/978-3-030-004 46-0_6

Matta G, Gjyli L (2016) Mercury, lead and arsenic: impact on environment and human health. J Chem Pharm Sci 9(2):718–725

McDonald K (2020) Food fraud. www.treetops.com. https://foodingredients.treetop.com/fruit-ing redients-blog/Post/Food-Fraud/.

Metcalfe DD, Sampson HA, Simon RA (eds) (2011) Food allergy: adverse reactions to foods and food additives. Blackwell Publishing Ltd, Maiden, Oxford, and Carlton

Mudgil D, Barak S, Khatkar BS (2011) Food additives in confectionery industry: an overview. Indian Food Pack 65(3):80–83

Murcia MA, Egea I, Romojaro F, Parras P, Jiménez AM, Martínez-Tomé M (2004) Antioxidant evaluation in dessert spices compared with common food additives. Influence of irradiation procedure. J Agric Food Chem 52, 7:1872–1881. https://doi.org/10.1021/jf0303114

Nigam PS, Luke JS (2016) Food additives: production of microbial pigments and their antioxidant properties. Curr Opin Food Sci 7:93–100. https://doi.org/10.1016/j.cofs.2016.02.004

Nithin CT, Chatterjee NS, Joshy CG, Yathavamoorthi R, Ananthanarayanan TR, Mathew S (2020) Source-dependent compositional changes in coconut flavored liquid smoke and its application in traditional Indian smoked fishery products. Food Addit Contam Part A 37(10):1610–1620. https://doi.org/10.1080/19440049.2020.1798030

Palaniyappan V, Nagalingam AK, Ranganathan HP, Kandhikuppam KB, Kothandam HP, Vasu S (2013) Antibiotics in South Indian coastal sea and farmed prawns (Penaeus monodon). Food Addit Contam Part B 6(3):196–199. https://doi.org/10.1080/19393210.2013.787555

Rather SA, Masoodi FA, Akhter R, Rather JA, Shiekh KA (2016) Advances in use of natural antioxidants as food additives for improving the oxidative stability of meat products. Madridge J Food Technol 1(1):10–17. https://doi.org/10.18689/mjft.2016-102

Saltmarsh M (ed) (2000) Essential guide to food additives. Leatherhead Publishing, LFRA Ltd., Leatherhead

Satyanarayana S, Sushruta K, Sarma GS, Srinivas N, Raju GS (2004) Antioxidant activity of the aqueous extracts of spicy food additives—evaluation and comparison with ascorbic acid in in vitro systems. J Herb Pharmacother 4(2):1–10. https://doi.org/10.1080/J157v04n02_01

Sharma A (2018) Evaluation of certain food additives and contaminants. Indian J Med Res 148(2):245–246

Silva MM, Lidon FC (2016) An overview on applications and side effects of antioxidant food additives. Emir J Food Agric 28:823–832

Singh N, Bajaj IK, Singh RP, Gujral HS (2003) Effect of different additives on mixograph and bread making properties of Indian wheat flour. J Food Eng 56(1):89–95. https://doi.org/10.1016/S0260-8774(02)00151-6

Singla RK, Dubey AK, Garg A, Sharma RK, Fiorino M, Ameen SM, Haddad MA, Al-Hiary M (2019) Natural polyphenols: chemical classification, definition of classes, subcategories, and structures. J AOAC Int 102(5):1397–1400. https://doi.org/10.1093/jaoac/102.5.1397

Soni KB, Lahiri M, Chackradeo P, Bhide SV, Kuttan R (1997) Protective effect of food additives on aflatoxin-induced mutagenicity and hepatocarcinogenicity. Cancer Lett 115(2):129–133. https://doi.org/10.1016/S0304-3835(97)04710-1

Chapter 5
Food Safety and Quality Control in Food Industry

Abbreviations

CFIA	Canadian Food Inspection Agency
CFSQP	Canadian Food Safety and Quality Program
CFDA	China Food and Drug Administration
CAC	Codex Alimentarius Commission
EFSA	European Food Safety Authority
F&B	Food and Beverage
FBO	Food Business Operator
FSSAI	Food Safety and Standards Authority of India
FSMS	Food Safety Management System
GRAS	Generally Recognized as Safe
GHP	Good Hygiene Practice
GMP	Good Manufacturing Practice
HACCP	Hazard Analysis and Critical Control Points
ISO	International Organisation for Standardization
QMS	Quality Management System
TQM	Total Quality Management
USFDA	United States Food and Drug Administration
WHO	World Health Organization

5.1 A General Discussion Concerning Food Safety

Around the world, an estimated 600 million—almost one in 10 people—fall ill after eating contaminated food each year (WHO 2020). Every nation has regulations to

© The Author(s), under exclusive license to Springer Nature Switzerland AG 2022
S. M. Varghese et al., *Trends in Food Chemistry, Nutrition and Technology in Indian Sub-Continent*, Chemistry of Foods,
https://doi.org/10.1007/978-3-031-06304-6_5

protect its people against unsafe practices in food production, and the existing agencies need to be equipped to change control (Khurana 2016). Good hygiene principles need to be regulated and enforced. In India, there is a need for enforcement of standards, and proper safety management. Public awareness programmes on food safety will increase the demand for quality and safe food in India.

Food Safety begins 'from Farm to Fork', with the initial production of raw materials in agricultural fields, industrial food processing, packing, transport, distribution, and final consumption. The concept of food safety is to implement food safety and quality systems throughout the entire food chain. Quality assurance ensures a set of standards for the products as per the specifications. Good Hygiene Practices (GHP) and Good Manufacturing Practices (GMP) ensure food safety by improving the hygienic conditions in the production of food, packing, storing, or distribution (Knaflewska and Pośpiech 2007). The role of the food and beverage operators, processors, and technologists is to prevent undesirable changes and to obtain desirable changes in food.

Food safety ensures that food is possibly free from all possible contaminants and hazards. A major challenge in the food industry is to motivate food handlers to apply the principles of Hazard Analysis and Critical Control Points (HACCP) (Kamboj et al. 2020). Food safety is threatened by a variety of foodborne diseases (Borchers et al. 2010). Foodborne illness is a major public health problem worldwide. The supply of safe and healthy food is crucial to preventing foodborne illness (Alemayehu et al. 2021). The types and amounts of microorganisms remaining in the food after various processing methods also affect durability.

Numerous chemical contaminants get into the human body through food from cultivation, processing, and cooking of foods. Pesticides, like organophosphates, applied to food crops may affect brain development. Many other chemicals used today are suspected carcinogens. Food contaminants also leach from the packaging or storage containers like phthalates and bisphenol-A that can affect the central nervous system and the male reproductive organs (Borchers et al. 2010).

It is essential to educate consumers about the relation between food and diseases and the importance of making proper food choices for consumption. Food producers, distributors, handlers, and vendors are completely responsible for this ambit, and food consumers should be always aware of food-safety-related implications. Government agencies must enforce food safety laws to safeguard public and individual health. The intimate collaboration between all the stakeholders will ultimately ensure adequate strategies for food safety, especially when speaking of cold-chain management. Differences in consumer preferences on food values ensure enhanced marketing strategies in the import and export food market (Fung et al. 2018; Gurudasani and Sheth 2009). Thus, we can understand that food safety corresponds to the integrity and wholesomeness at all stages of production, processing, and distribution of food, including rules aimed at ensuring fair practices in trade and protecting consumer interests and information, the manufacture and use of materials and articles intended to come into contact with foods.

5.2 Quality Control and Quality Management Systems

Quality control ensures product quality through efficient management by training employees, creating benchmarks for product quality, and testing products to check for statistically significant variations. A significant aspect of quality control is the establishment of well-defined control points. Quality Management System (QMS) is the commitment attained through Standard Operating Procedures) that imparts consistent and compliant outcomes (Elahi 2018). GMP, HACCP, and International Organisation for Standardization (ISO) enable food quality and safety assurance (Rotaru et al. 2005; Shafiee and Drabu 2017).

The HACCP approach was developed to ensure the food safety of astronauts by the Pillsbury Company for the National Aeronautics and Space Administration and the United States Army laboratories at Natick. Today, HACCP is an internationally accepted concept for assuring food safety through the identification and control of biological, physical, and chemical hazards in the food and beverage sector (Slatter 2003). HACCP system is a preventive system of hazard control rather than a reactive one. Active managerial control practices are developed to ensure that the hazards are eliminated or minimized. HACCP enables to attain specifications and customer expectations, ensuring food safety. In relation to food, GMP assures food safety through vigilant measures at the source product design and process control (Frestedt 2017).

Codex Alimentarius is translated from Latin, a 'food code'. It comprises a series of general and specific food safety standards that have been formulated with the objective of protecting consumer health and ensuring fair practices in the food trade. Food must be safe to eat, of good quality, and should not carry disease-causing organisms that could harm animals or plants in importing countries. Codex Alimentarius is run by the Codex Alimentarius Commission, which is an intergovernmental body for drafting standards, adopted by the Codex Commission.[1]

ISO's food safety management standards help organizations identify and control food safety hazards, at the same time as working together with other ISO management standards, such as ISO 9001. There is also ISO 22000, a different standard that is specifically applicable to all food producer systems. ISO 22000 provides a layer of reassurance within the global food supply chain, helping products cross borders and bringing people food that they can trust.[2]

A hybrid approach among ISO 9001, QMS, and HACCP is necessary for improving food safety. On the other side, the application of good manufacturing practices, HACCP, and ISO 9001:2000 separately is not a good strategy. ISO 22000—also known as Food Safety Management System (FSMS)—is an international auditable standard. ISO 22000 incorporates critical control point and hazard analysis systems

---

[1] The interested Reader can find more information at the following web site address: https://www.fao.org/fao-who-codexalimentarius/en/.

[2] The interested Reader can find more information at the following web site address: https://www.iso.org/iso-22000-food-safety-management.html.

in the more improved form to produce more effective auditable FSMS, and the final aim is to assure reliability, food quality, and food safety (Panghal et al. 2018).

Standards ensure and enhance transparency in the development of food quality and safety procedures, thus helping to upgrade and update food safety systems to evolve effective Total Quality Management (TQM) system. TQM is a managerial approach for organizations focused on quality (Fayaz et al. 2020). In the future, technological progress may provide economic and efficient tools to predict and solve food safety issues (Khan et al. 2021).

5.3 Food Hazards in Terms of Physical, Chemical, or Biological Contaminants

Hazard is defined as 'a biological, chemical or physical agent in food, including allergens, or a condition of food with the potential to cause an adverse health effect' (CAC 1997). Unsafe food containing different pathogens and chemical dangerous substances are estimated many major human illnesses (WHO 2020). Food safety involves strategies and activities aimed to protect foods from biological, chemical, physical, and allergenic hazards.

Contamination can be intentional or contamination by chemicals such as pesticide residues, physical contaminants such as stones, metals, grits, or biological agents (Connolly et al. 2016; Michaelidou and Hassan 2008).

Toxins produced by moulds (mycotoxins) can be fatal as they have teratogenic, mutagenic, or carcinogenic effects. Amino acids of animal origin are broken down by bacteria to produce toxins like histamines and other biogenic amines that can cause foodborne illness. Environmental contaminants, metals, pesticide and drug residues, animals, birds, and rodents can cause contamination of food.

Food and water can transmit diseases and food infections, through food intoxication. Parasites and helminths can cause foodborne illnesses. With reference to specific microbial menaces of main importance, *Listeria monocytogenes* is an important foodborne pathogen. It is a causative agent of a severe infection that primarily affects immunocompromised people, pregnant women, and occasionally healthy people. Intrauterine infection of the foetus resulted in a dearth or an actually ill infant with a septic disseminated from listeriosis. Faecal pollution in food and water is the major cause of disease outbreaks, a significant number of novel genetic markers of faecal indicator bacteria have been identified (Zheng and Shen 2018).

A chemical hazard can cause immediate or long-term effects on consumption. Chemical hazards are common due to accidental spillage of permitted additives or Generally Recognized as Safe (GRAS) chemicals. GRAS may be dangerous especially if they exceed permissible levels. Usage of non-permitted additives is common in developing countries, where enforcement is delayed or neglected.

Any extraneous object or foreign matter in a food that causes illness or injury to a person on consumption is a physical hazard. Systems should be in place to prevent contamination of foods by foreign bodies such as glass, metal shards, etc., and unallowed/extraneous chemicals. In manufacturing and processing, suitable detection or screening devices should be used wherever necessary. The possible source of physical hazards may be identified by untrained operators (employees). The presence of non-food-related parts or contaminant materials can cause injury, wounds, and infections in the throat and intestine. Hence, due importance must be given to personnel hygiene, protective suits, gloves, footwear, head caps, facial masks, and proper use of jewellery.

5.4 Intrinsic and Extrinsic Factors Affecting the Growth of Microbes in Food

Improper food processing methods enable microorganisms to grow in food causing deterioration in food quality. The degree of deterioration is based on the amount and spreading of toxins produced by microbes during their metabolic activity. There are microbes that bring about favourable and unfavourable changes in food. Intrinsic, extrinsic, and implicit factors affect the growth of microbes in food. Intrinsic factors are those related to food, i.e. nutrient content, water activity, pH value, redox potential, antimicrobial substances, and mechanical barriers to microbial invasion. On the other side, extrinsic factors are related to the environment, i.e. storage temperature, composition of gases and relative humidity. Finally, implicit factors concern interactions between the same or different types of microbes in food (Hamad 2012; Preetha and Narayanan 2020).

Microorganisms such as moulds, yeasts, and bacteria can grow in food and cause spoilage. Bacterial spreading by spore-producing life forms results in the spoilage of foods. Good examples are *C. botulinum, C. perfringens, B. cereus*, and *B. anthracis*. Bacterial spores can survive in chemicals, alcoholic compounds and high temperatures of 100–110 °C. However, spores will start reproducing in a favourable environment. Moulds require sufficient moisture, air, and temperature to grow: they can be found and visibly detected on various foodstuffs. Favourable conditions include refrigerators, improperly processed foods, and dry or semi-dry foods with very low moisture. *Aspergillus (A. flavus* and *A. parasiticus), Rhizopus*, and *Penicillium* species are common foodborne pathogens. Yeasts also spoil the food to a great extent and can resist conditions like alkalinity, acidity, dehydration, and very low temperatures.

Foodborne pathogens are a real threat to human health and the economy. Pathogenic bacteria (*B. cereus, Campylobacter jejuni, C. botulinum, C. perfringens, Cronobacter sakazakii, Escherichia coli, L. monocytogenes, Salmonella* spp, *Shigella* spp, *Staphylococcus aureus, Vibrio* spp, and *Yersinia enterocolitica*), viruses (Hepatitis A and Noroviruses), and parasites (*Cyclospora cayetanensis, Toxoplasma gondii*, and *Trichinella spiralis*) play significant roles in food spoilage (Bello 2021; Bintsis 2017; Rawath 2015).

Food spoilage can be defined as any unacceptable sensory change in food that leads to deterioration of quality. The rate of contamination in food is proportional to unsanitary methods of handling them from harvesting to transportation and storage (Bello 2021).

Food intoxication caused by *S. aureus* and *Salmonella* spp results in a high rate of morbidity and mortality in developing countries. Consequently, adequate foodborne disease surveillance is highly recommended (Sudershan et al. 2014).

5.5 Regulatory Agencies

International food-safety-related legislations should reflect new knowledge, and technical innovations, addressing current issues; they should also develop guidelines for facilitating fair trade. National regulatory guidelines of each country monitor activities of food business operators, like the United States Food and Drug Administration (USFDA), European Food Safety Authority (EFSA), China Food and Drug Administration (CFDA), and Food Safety and Standards Authority of India (FSSAI). Food manufacturers are required to meet the given standards of quality and food safety and need to get certified to ensure the defence mechanism against many foodborne diseases.

USFDA—established in 1906—has created a safety assessment panel including a team of expert chemists and toxicologists. The USFDA Food Code is a model for safeguarding public health and ensuring food is unadulterated and honestly presented when offered to the consumer. It represents the FDA's best advice for a uniform system of provisions that address the safety and protection of food offered at retail and in food service (USFDA 2017).

EFSA—established in 2002—provides scientific advice and communicates on existing and emerging risks associated with the food chain. EFSA issues advice on existing and emerging food risks. Food and feed safety, nutrition, animal health and welfare, plant protection, and plant health are the major areas of concern for EFSA.

CFDA—established in 2013—drafts laws and regulations on the safety management of food, health food and cosmetics, policy development, and related implementation. It also ensures safety management of food, with foreign governments and international organizations.

FSSAI has been established under the Food Safety and Standards act in 2006. FSSAI's main concerns include ensuring food manufacture, storage, distribution, sale, and import with the aim of assuring the availability of safe and wholesome food for human consumption.

Food Safety Enhancement Program (FSEP) developed by the Canadian Food Inspection Agency (CFIA) follows the HACCP-based food safety systems in federally registered agricultural food processing establishments. The CFIA and the food sector have developed generic models for many commodities and a mandatory Quality Management Program for fish and seafood. Today, the Canadian approach to food safety is complemented by the Canadian Food Safety and Quality Program (CFSQP) headed up under the Agricultural Policy Framework.

In Australia, Food Standards Australia-New Zealand developed food standards that cover the entire food supply chain, from paddock to plate. Food safety regulation in Russia involves a number of federal government bodies, each with their own area of specialization. These include the National Body of Sanitary Control, Ministry of Agriculture, Agency of Technical Regulation, State Trading Inspection, Ministry of Economic Development, and State Grain Inspection. Each is involved in traceability, safety, and hygiene for the food market in Russia.

The situation is really different in developing countries because of lacking of technology and awareness. The main part of food producers is on a mini-, micro-, and small-scale dimensions. The implementation of food safety legislation in these areas needs further modifications to ensure food safety.

5.6 The Problem of Food Safety in the Industry of Foods and Beverages. The Viewpoint of Food Business Operators

Food safety and quality control ensure the food is safe for human consumption. This reflection sums up the perception of food safety quite well as perceived by the stakeholders of the entire food chain. On the other hand, it is not so easy to put a similar concept into practice when we consider that the food chain contains not only stakeholders producing goods and services, from primary production to final distribution, passing through all the intermediate phases including intermediate storage, international logistics, etc. In fact, many other stakeholders—the Official Authorities, Country by Country; the Regulatory Bodies; the scientific Associations that deal with regulations and re-evaluation of regulatory or voluntary standards over time—should be considered. Finally, the role of the final consumer of the final product and/or service should be considered with attention. The problem is here. On the one hand, proper food supply chain management benefits both the manufactures—and, more in general, the Food Business Operators (FBO)—and consumers. On the other hand, it is necessary to ensure that reliable quality assurance and quality control processes are implemented to supply safe food and beverage products to the consumer. National Regulatory Guidelines in many countries monitor FBO activities, such as the United States Food and Drug Administration, the European Food Safety Authority, the China Food and Drug Administration, and the Food Safety and Standards Authority of India.

On these bases, the present situation of the industry of Foods and Beverages (F&B) is as follows: FBO—including also F&B manufacturers—are required to meet mandatory or voluntary standards of quality and food safety, and need to get certified to ensure the defence mechanism against many foodborne diseases (Bai et al. 2007; Bilska and Kowalski 2014; Bosona and Gebresenbet 2013; Botonaki et al. 2006). These arguments are debated for decades (Barbieri et al. 2014; Barone et al. 2014; Brunazzi et al. 2014; Haddad et al. 2021a, b; Jisung and Lee 2021; Laganà et al. 2015; Osama 2018; Urmila et al. 2017). However, what about the real perception of quality control and effectiveness? In fact, the difference between real-time checks and the evidence of already carried out checks should be evaluated.

The viewpoint of FBO is—or should be—considered with attention because these players are 'part of the game'. In fact, the main—or one of the main problems—in the evaluation of food-oriented quality systems and also plans devoted to the production of safe foods and beverages—also named HACCP plans because of the well-known acronym for Hazard Analysis and Critical Control Points (HACCP)—is the correct representation of performed controls in all the steps of a food chain if the final recipient of these information is not an FBO. In fact, each FBO-related player has to take into account, implement, carry on, and finally re-examine the results of its action during food-related production processes, including also other important and critical steps (storage, logistics, etc.). On the other side, the mass of information stored in digitalized documents or written on paper sheets may become incomprehensible for the common Consumer, and—in the worst situation—for the industrial customer or the Official Authority performing controls. Consequently, the viewpoint of each player performing quality controls with some food safety-related importance is to simplify operations within a reasonable period of time. Consequently, all procedures and instructions in a player-oriented manual, instruction, or procedure, are—or should tend to be—simple enough. As a result, each recording document tends to describe controls in a simplified manner, taking into account basic descriptions of each control without legends, and with some sort of tacit agreement between the writer of these instructions and simple operators. This situation often causes misunderstandings between audited companies/directors/operators on the one side, and auditors (private or official inspectors) on the other side. In conclusion, the development of efficient/reliable food safety plans based on quality controls should aim at the comprehension of results. After all, voluntary standards such as the International Featured Standards (IFS) Food ask has different objectives, including the necessity of providing evidence of commitment. In other terms, quality control rules have to be:

(a)   Easily applicable to operators, and
(b)   Easily understandable by auditors and other interested stakeholders without need of excessive clarifications.

5.7   The Problem of Food Safety in the Industry of Foods and Beverages. The Viewpoint of Non-FBO Players. Towards a Difficult 'Translation'

On the basis of the above-discussed reflections, the comprehension of food-related procedures, instructions, plans, manuals, and—above all—working documents of registration into a food-related plant or similar organization is critical. In other words, the use of quality control-based procedures should be not excessively complicated for operators, but the result of controls has to be necessarily understandable by non-FBO players.

The above-mentioned reflection is not easy, especially if the writer of food safety plans acts simply as a person able to reason only in terms of production, supply management, logistics, and so on. One of the main mistakes in this ambit is simplified in Fig. 5.1, where a 'simple' record sheet shows a succession of thermal records along a single production line. The record sheet easily shows all operations made by operators during a specified time period. However, this succession or series of data can be extremely readable and understandable by operators only, whereas the same series of numbers can be absolutely incomprehensible by another person, including also specialized auditors. The reason is simple enough: the operator reasons/is accustomed to reason in terms of controls along a single line and a specified temporal period (Fig. 5.1). On the other hand, the same operator does not include important information in the same sheet concerning the identification of the real subject of controls: the F&B product.

RECORD SHEET No 1			
Date	03-02-2022	**Responsible Operator**	
Line	5.2	* * * * * * * * * *	

Hour >>		Control>>	Temperature °C
08:00			81
09:01			79
10:05			78
11:10			82
12:00			84
13:05			85
14:07			79
15:11			80
16:02			78
17:00			77
18:02			76

All product identification data are reported in Record Sheet No 2

Fig. 5.1 A simple record (electronic) sheet showing quality controls in a food industry… from the viewpoint of FBO players. There are different thermal data recorded at different times on a single production line. Unfortunately, the identification of food products—in terms of lot, expiration date, name, brand, etc.—is not evident here (they are explicitly placed in a different worksheet, as declared…). This document is highly readable and reliable for FBO players. Unfortunately, non-FBO players do have not the 'right interpretation key'…

Actually, the cause of this mistake cannot be ascribed to the operator: in general, the critical information concerning product traceability is probably written/recorded in a different section of the same food-related record, which is composed of different sheets or digitalized documents. After all, there is some limitation to the amount of data that can be stored on a single sheet, including also electronic sheets.

The solution is to reason in terms of non-FBO players. In other words, all evidence concerning food safety have a clear and often mentioned subject: the food product, in terms of type, prize, weight, lot, expiration date, etc. Consequently, the same information is shown in Fig. 5.1—surely real, readable, and certainly reliable when speaking of quality-control effectiveness—should be 'translated' to the final recipient of the information who is not aware of the mass of tacit knowledge behind each step along the line of food production, storage, handling, and final distribution. In other terms, the same information could/should be supplied to the interested non-FBO player in terms of the exact identification of the questioned food product. Figure 5.2 shows a simple 'translation' of the same situation displayed in Fig. 5.1 with one difference: all data are related to one simple reference only, while the remaining data—concerning the complexity of a whole organized summary of operations along a specified temporal period, and on the same production line—remain 'masked'. In fact, each non-FBO player is only interested in the food-safety management of one product only, while the sum or all operations can be difficultly understandable. Actually, the general opinion of auditors in this ambit is not favourable if they are forced to

RECORD SHEET No 1

| Date | 03/02/2022 | | **Responsible Operator** | | | |
| Line | 5.2 | | * * * * * * * * * | | | |

Product >> ▾	Lot/TBD ▾	Controls ▾	▾	▾	Temperature ▾	▾
		VS Hour	08:00	09:01	10:05 11:10	12:00
Cake # 3	279/02.02.24		81	*	* *	*
Pizza Mon	279/02.02.23		*	79	78 82	*
Intermediate # 5	279/01.03.22		*	*	* *	78
RockFood	279/02.02.24					

Fig. 5.2 A simple record (electronic) sheet showing quality controls in a food industry… from the viewpoint of non-FBO Players. Differently from Fig. 5.1, there are only several information concerning all product. These data can be filtered for one food product only, so that the immediate question can have immediate (and readily understandable) answers…

RECORD SHEET No 1

Date	03/02/2022
Line	5.2

Responsible Operator
****

Product >>	Lot/TBD	Control				Temperature	
Pizza Mon	279/02.02.23		*	79	78	82	*

Fig. 5.3 A simple record (electronic) sheet showing quality controls in a food industry… from the viewpoint of non-FBO Players. This sheet derives from the record sheet of Fig. 5.2: present data have been filtered with the aim of offering only information concerning one product only, so that the immediate question can have immediate (and readily understandable) answers…

control a mass of apparently incoherent data. The reason for this evaluation depends on the nature of the management of a reliable food safety plan or quality system. It can be challenging enough when speaking of a complete industrial environment. Consequently, the only strategy able to rapidly understand the validity or unreliability of such a system is to consider only a small dataset related to one product only. For these reasons, the management of food safety systems needs strongly to be:

(a) Implemented in an easy way when speaking of material execution, and
(b) Translated and offered to non-FBO players in a readable/comprehensible way, using only one interpretation key: the filtration of all available data on the basis of a simple product identification (as shown in Fig. 5.3, derived from Fig. 5.2).

In our opinion, the next evolution of food-quality systems should aim at the easy conceptualization and translation of operative records/results when offered to the final listener, as shown by means of the comparison of Figs. 5.1 and 5.2. Otherwise, the real risk of food safety systems is to create records that can be understood only by the creator/recorder, while the remaining part of interested stakeholders—which are the real recipients of such challenging efforts—cannot probably comprehend similar efforts, and the analytical significance of records. The description of these data—raw data such as time, temperature, viscosity, chemical/microbiological evaluations, etc.—cannot help the professional because each control has its importance in the strict and limited ambit of the step(s) assigned in the food safety plan, and in the geographical ambit of the peculiar industry. In other terms, a determined result can have a low importance in one particular ambit/industry/ date, and other (very) different weights in other ambits/companies/temporal periods.

5.8 Conclusions

Food industry is one of the fastest-growing industries in India and around the world. Food and beverage business operators are required to meet the given standards of quality and food safety, and need to get certified to ensure the defence mechanism against many foodborne diseases Enforcement of GMP, GHP, HACCP, and FSMS ensures complete food safety.

Food safety should be a public health priority, for the food producers and suppliers, throughout the food chain to act responsibly and provide safe food to consumers. Though food contamination can occur at any stage of the manufacturing or distribution process. The food and beverage producers at home, industry, or street markets are responsible for total quality management. Improperly prepared or handled food is the cause of majority of the foodborne diseases. It is a joint responsibility of each of us including the appellate authorities and governments to work together to enforce laws that ensure Food Safety and Quality Control in the Food Industry.

References

Alemayehu T, Aderaw Z, Giza M, Diress G (2021) Food safety knowledge, handling practices and associated factors among food handlers working in food establishments in Debre Markos Town, Northwest Ethiopia, 2020: institution-based cross-sectional study. Risk Manag Healthc Policy 14:1155–1163. https://doi.org/10.2147/rmhp.s295974

Bai L, Ma C, Gong S, Yang Y (2007) Food safety assurance systems in China. Food Control 18(5):480–484. https://doi.org/10.1016/j.foodcont.2005.12.005

Barbieri G, Barone C, Bhagat A, Caruso G, Conley ZR, Parisi S (2014) The influence of chemistry on new foods and traditional products. Springer International Publishing, Heidelberg, Cham. https://doi.org/10.1007/978-3-319-11358-6

Barone C, Bolzoni L, Caruso G, Montanari A, Parisi S, Steinka I (2014) Food packaging hygiene. Springer International Publishing, Cham. https://doi.org/10.1007/978-3-319-14827-4

Bello MI (2021) Isolation and characterization of food spoilage microorganisms at Jimeta by-pass market. Direct Res J Pub Health Environ Technol 6:67–70. https://doi.org/10.26765/DRJPHE T903712653

Bilska A, Kowalski R (2014) Food quality and safety management. Sci J Log 10(3):351–361

Bintsis T (2017) Foodborne pathogens. AIMS Microbiol 3:3:529–563. https://doi.org/10.3934/mic robiol.2017.3.529

Borchers A, Teuber SS, Keen CL, Gershwin ME (2010) Food safety. Clin Rev Allergy Immunol 39(2):95–141. https://doi.org/10.1007/s12016-009-8176-4

Bosona T, Gebresenbet G (2013) Food traceability as an integral part of logistics management in food and agricultural supply chain. Food Control 33(1):32–48. https://doi.org/10.1016/j.foo dcont.2013.02.004

Botonaki A, Polymeros K, Tsakiridou E, Mattas K (2006) The role of food quality certification on consumers' food choices. Brit Food J 108(2):77–90. https://doi.org/10.1108/000707006106 44906

Brunazzi G, Parisi S, Pereno A (2014) The importance of packaging design for the chemistry of food products. SpringerBriefs in Chemistry of Foods, Springer International Publishing, Cham. https://doi.org/10.1007/978-3-319-08452-7

CAC (1997) Joint FAO/WHO Food Standards Programme, Codex Committee on Food Hygiene. Vol. 1B-1997 suppl. Hazard Analysis and Critical Control Point (HACCP) System and guidelines for its application, Annex to CAC/RCP 1–1969, Rev. 3 1997. Codex Alimentarius Commission (CAC), Rome

Connolly A, Luo LS, Connolly KP (2016) Global insights into essential elements food safety: the Chinese example. J Food Prod Mark 22(5):584–595. https://doi.org/10.1080/10454446.2016.1141144

Elahi B (ed) (2018) Safety risk management for medical devices. Academic Press, Cambridge. https://doi.org/10.1016/B978-0-12-813098-8.00037-4

Fayaz H, Kumar A, Kousar F, Sharma S, Kumar S (2020) Application of total quality management to ensure food quality in food industry. J Anim Res 10(3):329–338. https://doi.org/10.30954/2277-940X.03.2020.1

Frestedt JL (2017) Hazard analysis and critical control points. In: Frestedt JL (ed) FDA warning letters about food products, pp 51–89. https://doi.org/10.1016/B978-0-12-805470-3.00003-X

Fung F, Wang H-S, Menon S (2018) Food safety in the 21st century. Biomed J 41(2):88–95. https://doi.org/10.1016/j.bj.2018.03.003

Gurudasani R, Sheth M (2009) Food safety knowledge and attitude of consumers of various food service establishments. J Food Saf 29(3):364–380. https://doi.org/10.1111/j.1745-4565.2009.00162.x

Haddad MA, Yamani MI, Abu-Romman SM, Obeidat M (2021a) Chemical profiles of selected Jordanian foods. SpringerBriefs in molecular science. Springer, Cham. https://doi.org/10.1007/978-3-030-79820-8

Haddad MA, Yamani MI, Da'san MMJ, Obeidat M, Abu-Romman SM, Parisi S (2021b) Food traceability in Jordan current perspectives. Springer International Publishing, Cham. https://doi.org/10.1007/978-3-030-66820-4

Hamad S (2012) Factors affecting the growth of microorganisms in food. In: Bhat R, Alias AK, Paliyath G (eds) Progress in food preservation. Wiley-Blackwell, Ltd., Chichester, pp 405–427. https://doi.org/10.1002/9781119962045.ch20

Jisung J, Lee E-K (2021) How do consumers' food values across countries lead to changes in the strategy of food supply-chain management? Foods 10(7):1523. https://doi.org/10.3390/foods10071523

Kamboj S, Gupta N, Bandral JD, Gandotra G, Anjum N (2020) Food safety and hygiene: a review. Int J Chem Stud 8(2):358–368. https://doi.org/10.22271/chemi.2020.v8.i2f.8794

Khan N, Ray RL, Sargani GR, Ihtisham M, Khayyam M, Ismail S (2021) Current progress and future prospects of agriculture technology: gateway to sustainable agriculture. Sustain 13(9):4883. https://doi.org/10.3390/su13094883

Khurana CG (2016) A study of food safety and hygiene in India. Int J Adv Res Innov Ideas Educ 2(2):169–175

Knaflewska J, Pośpiech E (2007) Quality assurance systems in food industry and health security of food. Acta Sci Pol Technol Aliment 6(2):75–84

Laganà P, Caruso G, Barone C, Caruso G, Parisi S, Melcarne L, Mazzù F, Delia AS (2015) Microbial toxins and related contamination in the food industry. Springer International Publishing, Cham. https://doi.org/10.1007/978-3-319-20559-5

Michaelidou N, Hassan LM (2008) The role of health consciousness, food safety concern and ethical identity on attitudes and intentions towards organic food. Int J Consum Stud 32(2):163–170. https://doi.org/10.1111/j.1470-6431.2007.00619.x

Osama OI (2018) Understanding the nature, physiology, taxonomy, diagnostic and the federal compliance guidelines for food-borne pathogen Listeria monocytogenes. In: Proceedings of the international conference on food safety & regulatory—water microbiology, water sustainability and reuse technologies, 03–04 Dec 2018, Chicago, USA. J Food Microbiol Saf Hyg 3. https://doi.org/10.4172/2476-2059-C4-016

Panghal A, Navnidhi C, Neelesh S, Sundeep J (2018) Role of food safety management systems in safe food production: a review. J Food Saf 38(4):e12464. https://doi.org/10.1111/jfs.12464

Preetha SS, Narayanan R (2020) Factors influencing the development of microbes in food. Shanlax
    Int J Arts Sci Humanit 7(3):57–77. https://doi.org/10.34293/sijash.v7i3.473
Rawath S (2015) Food spoilage: microorganisms and their prevention. As J Plant Sci Res 5(4):47–56
Rotaru G, Sava N, Borda D, Stanciu S (2005) Food quality and safety management systems: a
    brief analysis of the individual and integrated approaches scientifical researches. Agroaliment
    Proc Technol 11:1:229–236. https://journal-of-agroalimentary.ro/admin/articole/40522L22_art
    icle-Rotaru_rev_IV.pdf. Accessed 26th Feb 2022
Shafiee MN, Drabu S (2017) Assessment of quality control systems in food processing units in
    Khunmoh Food Park Kashmir, India. Indian Hortic J 7(1):79–84. 403-16-IHJ-2911-16
Slatter J (2003) Hazard analysis critical control point. In: Benjamin Caballero B (ed) Encyclopedia
    of food sciences and nutrition, 2nd edn. Academic Press, Cambridge, pp 3023–3028. https://doi.
    org/10.1016/B0-12-227055-X/00580-0
Sudershan RV, Kumar RN, Kashinath L, Bhaskar V, Polasa K (2014) Foodborne infections and
    intoxications in Hyderabad India. Epidemiol Res Int 2014:942961. https://doi.org/10.1155/2014/
    942961
Urmila M, Nagar P, Maan S, Kaur K (2017) A growth of different types of microorganism, intrinsic
    and extrinsic factors of microorganism and their affects in food: a review. Int J Curr Microbiol
    App Sci 6(1):290–298. https://doi.org/10.20546/ijcmas.2017.601.035
USFDA (2017) Food code 2017. United States Food and Drug Administration (USFDA),
    Washington, DC. https://www.fda.gov/food/fda-food-code/food-code-2017. Accessed 26th Feb
    2022
WHO (2020) Food safety—key facts. World Health Organization (WHO), Geneva. https://www.
    who.int/news-room/fact-sheets/detail/food-safety. Accessed 26th Feb 2022
Zheng G, Shen Z (2018) Host-specific genetic markers of fecal bacteria for fecal source tracking
    in food and water. J Food Microbiol Saf Hyg 3:135. https://doi.org/10.4172/2476-2059.1000135

Chapter 6
Food Preservation

Abbreviations

GRAS   Generally Regarded as Safe

6.1 Introduction

At present, the main challenges for the modern industry of foods and beverages—including related economic cycles—are innovation, sustainability, and safety. In general, it can be affirmed that innovation aims at maintaining technological processes at the highest possible level. On the other side, food safety—normally considered the prioritized factor of the food production and preservation industry—should be considered as an ideal macro-container including innovation and process sustainability. In other terms, the quality of foods and beverages may be not an absolute value, while food safety is a mandatory and absolute requisite. Unfortunately, difficulties are occurring with a notable frequency at present when speaking of food regulatory compliance (Rahman 2020, also because food and beverage science corresponds to a multidisciplinary approach (Leistner 2000; Prokopov and Tanchev 2007)). This chapter would give some useful advices and information concerning modern technologies and food quality and safety.

As mentioned above, preservation systems should be part of a multidisciplinary effort concerning growing, harvesting, processing, packaging, and distribution along the food supply chain. Basically, each food or beverage with prolonged shelf life and correlated technological/sensorial/microbial/chemical/safety features would be highly desirable, even if the First Law of Food Degradation by Parisi establishes a critical limit: there is always some type of food/beverage modifications, and there is always a termination of desired durabilities (when defined). What about properties or

© The Author(s), under exclusive license to Springer Nature Switzerland AG 2022
S. M. Varghese et al., *Trends in Food Chemistry, Nutrition and Technology in Indian Sub-Continent*, Chemistry of Foods,
https://doi.org/10.1007/978-3-031-06304-6_6

characteristics which should/could be preserved (Rahman 2020)? Any property can be good or bad features depending on the product, the use, and the final consumer).

In the present scenario, food preservation is one of the main technological indicators in the modern science of foods and beverages, also with reference to marketing approaches and the prompt availability of certain 'off-season' fruits and vegetables (Seervi et al. 2014). Inadequate production and surplus amounts of the same food commodity in different places and seasons should be a critical problem, especially in pandemics times. Additionally, foods have a certain 'index of perishability', from high to moderate or low decay, and this situation has to be taken into account when speaking of preservation techniques able to save seasonal foods intact for later use (Aluga and Kabwe 2016).

The ambit of food preservation is really important in developing Nations, and India is a good example. Food stuffs such as vegetable materials, grains, and various fruits have remarkable importance in the hilly terrain of Himachal Pradesh (Seervi et al. 2014): in detail, the northern state area of Himachal Pradesh experiences huge climatic variations, such as other countries where hot and humid tropical conditions are alternated to cold and alpine conditions along the year. Moreover, some areas of this state may remain isolated due to harsh weather conditions and landslides (e.g., Spiti Valley); as a result, food accessibility may become a critical problem (Singh et al. 1997). Furthermore, cultivation happens only in selected months of the year depending on the geographical location. As a result (Aluga and Kabwe 2016):

(1)   A remarkable amount of fruits and vegetables may be available at cheap conditions in certain periods of the year, whereas unavailability has to be predicted in other months.

(2)   On these bases, preservation and storage of food for later consumption is certainly a need for interested populations.

Agriculture in Himachal Pradesh contributes nearly 45% to the net state domestic product. It is the main source of income as well as employment in the region. About 93% of the state population depends directly upon agriculture, which provides direct employment to 71% of its people (Economic Survey 2009–2010). As a consequence, cultivated crops should remain usable for long periods using proper techniques for later use. It has to be noted that a marketable supply of food products is maintained from Himachal Pradesh to other parts of India and the world (Seervi et al. 2014).

Food preservation is the process to handle and treat a food in order to control its spoilage by stopping the attack and growth of food borne diseases causing microbes; in this way, chemical oxidation of lipids, also named 'food rancidity', should be kept under control, also maintaining reasonably good nutritional profiles and sensorial features (Abdulmumeen et al. 2012).

Actually, food preservation systems are apparently one of the most important strategies able to overcome inappropriate planning in agricultural activities, in addition to the possibility of dietary variations for the general population (Rahman 2020). By means of these techniques, it may be expected that improved nutritional, functional, convenience, and sensory properties are obtained and kept reasonably constant over time, when speaking of modern foods and beverages. Moreover, consumer demand for healthier, sapid, various, and more convenient foods has to be taken into account (Rahman 2020), even in underdeveloped countries where the main dish corresponds or is related to a specific cereal (i.e., rice or wheat).

6.2 Principles of Food Preservation

Basically, food preservation should contrast microbial spoilage, also protecting the food/beverage from chemical alterations (Khan 2015; Mukhopadhyay et al. 2017). Technology has obtained interesting results in recent times. In fact, in earlier days, ice was used to preserve foods because of the well-known thermal control of microbial spoilage. Thus, the very low temperatures became an efficient method for preventing food spoilage. Let us now list the principles of food preservation:

(1) Removal of microorganisms or microbial inactivation. Technologies under these names may be carried out by removing air and water (in terms of inner moisture), lowering or increasing temperature, increasing the concentration of salt or sugar or acid in foods (Kautkar and Raj 2019). Water removal (until the most part of the inner moisture is eliminated) from leaves is critical when speaking of preservation of green leafy vegetables, and microbial spoilage is heavily inhibited in this way. This process is different from enzymatic inactivation

(2) Enzymatic inactivation. Microbial enzymes found in foods can be inactivated by modifying their conditions such as temperature and moisture. A useful example is 'blanching' (foods are boiled in hot water for some minutes), and the aim is to inactivate microbial enzymes by thermal augment rather than removing water (Chemat et al. 2017)

(3) Removal of insects, worms, and rats. In other words, food storage in dry, air-tight containers should preserve edible commodities from attacks by insects, worms, or rats.

6.3 Food Preservation Methods

Based on the mode of action, the major food preservation techniques can be categorized in the following ways (Fig. 6.1):

(a) Inhibition of chemical deterioration and microbial growth;

(b) Direct microbial or enzymatic inactivation (yeasts and moulds are not 'bacteria'; however, their inhibition is obtained by means of technological systems under the same name);

(c) Active prevention against re-contamination before and after processing (Gould 2000; Prokopov and Tanchev 2007; Rahman 2020).

Food Preservation in Practice today...

Fig. 6.1 Modern techniques concerning food preservation

While the currently used traditional preservation procedures continue in one or more of these three ways, great efforts have been recently observed when speaking of qualitative amelioration of food products because of two requests: (i) reduction of visible defects associated with food preservation (if treated products are compared with the original food), and (ii) desire for more natural foods (Rahman 2020).

Basically, preservation starts when the harvested foods are separated from the medium of immediate growth (plant, soil, or water) or meat from the animal after slaughter, or milk from the normal secretion of mammalian glands (Dabasso 2020; Rahman 2020). In other terms, food commodities have not been subjected to any treatment apart from cleaning and size grading in the case of vegetable materials (Rahman 2020). Postharvest technology is concerned with handling, preservation, and storage of harvested foods, and maintaining their original integrity, freshness, and quality, mainly by means of one or more of the following approaches (Rahman 2020):

(a)   Efficient control of the environmental atmosphere;
(b)   Implementation of adequate packaging/storage/transport systems;
(c)   Use of physical treatments such as curing, precooling, temperature treatments, cleaning, and waxing;
(d)   Use of chemical approaches against contamination, such as disinfection, fumigation, and dipping.

6.4   Food Preservation and Traditional Approaches

The discussion concerning traditional systems of food preservation includes many different techniques, which are briefly summarized as follows (Kumar 2019; Rahman

2020). The most known and used systems at present are certainly (Abdulmumeen et al. 2012; Hernández-Cortez et al. 2017; Radziejewska-Kubzdela et al. 2014; Tucker and Featherstone 2021):

(a) Refrigeration (storage between 4 and 10 °C);
(b) Boiling (heating at 100 °C in water);
(c) Sugar addition (hypertonic food matrices may notably inhibit microbial spreading);
(d) Pickling (use of vinegar or vegetable oils as liquid media for preserving vegetables and fruits);
(e) Canning (food cooking and sterilization or pasteurization into tin cans, jars, etc.);
(f) Fermentative processes using selected bacteria;
(g) Addition of food-grade chemicals (sugar, salt, benzoates, different preservatives, etc.). Several of these substances are briefly discussed in Chap. 4. It has to be considered that all of these allowed chemicals have been deemed 'Generally Regarded As Safe' (GRAS) in specified amounts that are specified (Del Olmo et al. 2017; Hurst et al. 2018).

On the other hand, some of these and other systems have interesting peculiarities. This chapter will briefly discuss two of these procedures in detail: curing and freezing approaches.

6.4.1 Curing

Reduction of inner water or moisture by means of systems such as osmosis is the main pillar when speaking of microbial inhibition via seasoning processes Moreover, curing processes can really enhance sapidity and flavour for certain foods on the condition that salt, sugar, nitrites, and/or nitrates are added. It has to be remembered that sodium chloride can reduce the extent of rancidity in certain foods (Kumar 2019; Rahman 2020).

6.4.2 Freezing

Basically, food commodities can be placed and maintained at −10 °C and lower temperatures with the aim of enhancing shelf-life values for long periods. In this ambit, microbial spreading is inhibited, but a counter-effect is that frozen foods should be heated at 75 °C before consumption. Some of these food commodities may be also stored between 0 and 10 °C in fridges for some time with the aim of avoiding excessive water losses. The most known methods are discussed here.

6.4.2.1   Sharp Freezing (Slow Freezing)

This technique, first used in 1861, involves freezing by the circulation of air, either naturally or with the aid of fans. The temperature may vary from −15 to −29 °C and freezing may take from 3 to 72 h (Desrosier and Tressler 1977). Ice crystals formed one large macrocrystal with consequent cellular breaking. The thawed tissue cannot regain its original water content. The first products to be sharply frozen were meat and butter. Nowadays, freezer rooms are maintained at −23 to −29 °C or even lower values, in contrast to the well-known temperature of −18 °C.

6.4.2.2   Quick Freezing

In this process, the food attains the temperature of maximum ice crystal formation (0 to −4 °C) in 30 min or less. Such a speed results in the formation of very small ice crystals and hence minimum disturbance of cell structure. Most foods are quickly frozen by one of the following three methods:

(a)   By direct immersion. Since liquids are good heat conductors, foods can be frozen rapidly by direct immersion in a liquid such as brine or sugar solution at a low temperature. Berries in sugar-solution-packed fruit juices and concentrates are frozen in this manner. The refrigeration medium must be edible and capable of remaining unfrozen at −18 °C and slightly below (Kandoran 2000; Noomhorm and Vongsawasdi 2004)

(b)   By indirect contact with refrigerant. Indirect freezing may be defined as freezing by contact of the product with a metal surface, which is itself cooled by freezing brine or other refrigerating media. This is an old method of freezing: the food or package is kept in contact with the passage through the refrigerant at −18 to −46 °C flows (Woolrich and Novak 1977)

(c)   By air blast. In this method, refrigerated air at −18 to −34 °C is blown across the material to be frozen. The advantages claimed for quick freezing over slow freezing (sharp freezing) are: (i) smaller (size) of formed ice crystals with consequent reduction of mechanical destruction of intact cells of the food, and (ii) the period for ice formation is shorter, therefore, with consequently reduced time for diffusion of soluble material and for separation of ice. Finally (iii), more rapid preservation of microbial growth and (iv) more rapid slowing down of enzyme action can be obtained (Desrosier and Tressler 1977).

Cryogenic Freezing

Although most foods retain their quality when quick-frozen by the above methods, a few of them require 'ultra-fast' freezing. Such materials are subjected to cryogenic freezing, which is defined as freezing at very low temperatures (below −60 °C) (Khadatkar et al. 2004). The refrigerant materials used at present in cryogenic freezing are liquid nitrogen and liquid carbon dioxide. In the former case, freezing may be achieved by immersion in the liquid, spraying of liquid, or circulation of its vapour over the product to be frozen.

Dehydro-Freezing and Freeze Drying

With reference to dehydro-freezing, partial dehydration is the initial part of the process. In case of some fruits and vegetables, about 50% of the moisture is removed by dehydration prior to freezing (Ahmed et al. 2016). This has been found to improve the quality of the food. Dehydration does not cause deterioration and dehydro-frozen foods are relatively more stable.

In relation to freeze drying, food is first frozen at -18 °C on trays in the lower chamber of a freeze drier and the frozen material is dried (in a two-step process) under a high vacuum (0.1 mm Hg) in the upper chamber. Direct sublimation of the ice takes place without passing through the intermediate liquid stage. The product is highly hygroscopic, excellent in taste and flavour, and can be reconstituted readily. Mango pulp, orange juice concentrate, passion fruit juice, and guava pulp are dehydrated by this method (Ravani and Joshi 2013).

6.5 Modern Methods for Food Preservation

6.5.1 Pasteurization

On the other side, current methods for food preservation tend to use different approaches, even from the viewpoint of consumers willing for minimally processed foods. Consequently, the following systems may be briefly remembered when speaking of the modern and large-scale food industry:

(a) Freeze-drying methods (moisture is removed under vacuum substantially involving ice sublimation);
(b) Pasteurization (different from drastic sterilization procedures);
(c) Vacuum-packaging. Oxidation of foodstuffs is notably inhibited by removing inner air into bags and containers. Naturally, anaerobic life forms can find here good survival and spreading opportunities (Blumenthal 1992; Reddy et al. 2018; World Health Organization 1981);
(d) Addition of chemical preservatives (Chap. 4);
(e) Pascalization (treatment of foodstuffs under heavy physical pressure);
(f) Bio-preservation by means of fermentative processes by lactic acid bacteria (Singh 2018).

6.6 Conclusions

The preservation and processing of food are not as simple or straightforward as it was in the past. It is now moving from an art to highly interdisciplinary science. A number of new preservation techniques are being developed to satisfy current demands of economic preservation and consumer satisfaction in nutritional and sensory aspects, convenience, safety, absence of chemical preservatives, price, and environmental safety (Rahman 2020). Understanding the effects on the food of each preservation method has, therefore, become critical in all aspects. At present, correct food preservation is critical (Majumdar et al. 2018). On the other side, economic factors and social responsibility may have their importance, and new systems may be needed in the modern industry of foods and vegetables.

References

Abdulmumeen HA, Risikat AN, Sururah AR (2012) Food: its preservatives, additives and applications. Int J Chem Biochem Sci 1:36–47

Ahmed I, Qazi IM, Jamal S (2016) Developments in osmotic dehydration technique for the preservation of fruits and vegetables. Innov Food Sci Emerg Technol 34:29–43. https://doi.org/10.1016/j.ifset.2016.01.003

Aluga M, Kabwe G (2016) Indigenous food processing, preservation and packaging technologies in Zambia. In: Proceedings of the indigenous knowledge systems symposium. Kisii University, Kisii, Kenya

Blumenthal D (1992) Food irradiation: toxic to bacteria, safe for humans. Department of Health and Human Services, Public Health Service, Food and Drug Administration, Office of Public Affairs, Wagshington, D.C.

Chemat F, Rombaut N, Meullemiestre A, Turk M, Perino S, Fabiano-Tixier AS, Abert-Vian M (2017) Review of green food processing techniques. Preservation, transformation, and extraction. Innov Food Sci Emerg Technol 41:357–377. https://doi.org/10.1016/j.ifset.2017.04.016

Dabasso BG (2020) Traditional meat processing knowledge, social-cultural values, nutritional quality and safety amongst the Borana women of Northern Kenya. Dissertation, Jomo Kenyatta University of Agriculture and Technology, Juja

Del Olmo A, Calzada J, Nuñez M (2017) Benzoic acid and its derivatives as naturally occurring compounds in foods and as additives: uses, exposure, and controversy. Crit Rev Food Sci Nutr 57(14):3084–3103. https://doi.org/10.1080/10408398.2015.1087964

Desrosier NW, Tressler DK (eds) (1977) Fundamentals of food freezing, 4th edn. AVI Publishing Co., Inc., Westport

Gould GW (2000) Preservation: past, present and future. Brit Med Bull 56(1):84–96. https://doi.org/10.1258/0007142001902996

Hernández-Cortez C, Palma-Martínez I, Gonzalez-Avila LU, Guerrero-Mandujano A, Solís RC, Castro-Escarpulli G (2017) Food poisoning caused by bacteria (food toxins). In: Malangu N (ed) Poisoning: from specific toxic agents to novel rapid and simplified techniques for analysis. InterchOpen, Rijeka

Hurst WJ, Finley JW, deMan JM (2018) Additives and contaminants. In: deMan JM, Finley JW, Hurst WJ, Lee CY (eds) Principles of food chemistry. Springer International Publishing, Cham, pp 527–565. https://doi.org/10.1007/978-3-319-63607-8_15

Kandoran MK (2000) Technological aspects of fish processing. Proceedings on the symposium on quality assurance in seafood processing. Society of Fisheries Technologists, Cochin, India, pp 41–53

Kautkar S, Raj R (2019) Preservation and shelf life enhancement of fruits and vegetables. Indian Farmer 6, 11:744–748. https://indianfarmer.net/assets/archieves/2019/NOVEMBER%202019.pdf#page=50. Accessed 8 Jan 2022

Khadatkar RM, Kumar S, Pattanayak SC (2004) Cryofreezing and cryofreezer. Cryog 44(9):661–678. https://doi.org/10.1016/j.cryogenics.2004.03.008

Khan MZA (2015) Protection of foods from microbes: a review on food preservation. Int J Pharma Biomed Res 2:13–18. http://www.ijpbr.net/form/2015%20Volume%202,%20issue%201/IJPBR-2015-2-1-13-18.pdf. Accessed 8th Jan 2022

Kumar A (2019) Food preservation: traditional and modern techniques. Acta Sci Nutr Health 3(12):45–49. https://doi.org/10.31080/ASNH.2019.03.0529

Leistner L (2000) Basic aspects of food preservation by hurdle technology. Int J Food Microbiol 55(1–3):181–186. https://doi.org/10.1016/S0168-1605(00)00161-6

Majumdar A, Pradhan N, Sadasivan J, Acharya A, Ojha N, Babu S, Bose S (2018) Food degradation and foodborne diseases: a microbial approach. In: Holban AM, Grumezescu AM (eds) Microbial contamination and food degradation. Elsevier, Amsterdam, pp 109–148. https://doi.org/10.1016/B978-0-12-811515-2.00005-6

Mukhopadhyay S, Ukuku DO, Juneja VK, Nayak B, Olanya M (2017) Principles of food preservation. In: Juneja V, Dwivedi H, Sofos J (eds) Microbial control and food preservation. Food microbiology and food safety. Springer, New York. https://doi.org/10.1007/978-1-4939-7556-3_2

Noomhorm A, Vongsawasdi P (2004) Freezing shellfish. In: Hui YH, Guerrero Legarretta I, Lim MH, Murrell KD, Nip W-K (eds) Handbook of frozen foods. Marcel Dekker, New York and Basel

Prokopov T, Tanchev S (2007) Methods of food preservation. In: McElhatton A, Marshall RJ (eds) Food safety. Springer, Boston. https://doi.org/10.1007/978-0-387-33957-3_1

Radziejewska-Kubzdela E, Biegańska-Marecik R, Kidoń M (2014) Applicability of vacuum impregnation to modify physico-chemical, sensory and nutritive characteristics of plant origin products—a review. Int J Mol Sci 15(9):16577–16610. https://doi.org/10.3390/ijms150916577

Rahman MS (2020) Handbook of food preservation, 3rd edn. CRC Press, Boca raton

Ravani A, Joshi D (2013) Mango and it's by product utilization—a review. Trends Post Harv Technol 1, 1:55–67. http://jakraya.com/journal/pdf/1-tphtArticle_4.pdf. Accessed 8 Jan 2022

Reddy SVR, Sharma RR, Gundewadi G (2018) Use of irradiation for postharvest disinfection of fruits and vegetables. In: Siddiqui MW (ed) Postharvest disinfection of fruits and vegetables. Elsevier, Amsterdam, pp 121–136. https://doi.org/10.1016/B978-0-12-812698-1.00006-6

Seervi P, Singhal H, Ingale S, Bhati A, Prazapati P (2014) Study of traditional methods of food preservation. Indian Institute of Technology, Mandi (India), ISTP (Group-7) final report, p 54

Singh VP (2018) Recent approaches in food bio-preservation—a review. Open Vet J 8(1):104–111. https://doi.org/10.4314/ovj.v8i1.16

Singh GS, Ram SC, Kuniyal JC (1997) Changing traditional land use patterns in the Great Himalayas: a case study of Lahaul Valley. J Environ Syst 25(2):195–211

Tucker G, Featherstone S (2021) Essentials of thermal processing. Wiley, Hoboken

Woolrich WR, Novak AF (1977) Refrigeration technology. In: Desrosier NW, Tressler DK (eds) Fundamentals of food freezing, 4th edn. AVI Publishing Co. Inc., Westport

World Health Organization (1981) Wholesomeness of irradiated food. Technical Report Series No. 659. World Health Organization, Geneva

Chapter 7
Popular Indian Weight Loss Diets—Pros and Cons

Abbreviations

BMI	Body mass index
low-carb	Carbohydrates-restricted
IER	Intermittent energy restriction
VLCD	Very low-calorie diet

7.1 Introduction to Obesity and Weight Loss

In India, the prevalence of obesity is rapidly increasing due to many factors such as globalization, urbanization, sedentary lifestyle, and dietary patterns (Joshi and Mohan 2018). The consequences are immense as it results in type-2 diabetes, cardiovascular disease, liver illness, arthritis, cancer, and many other associated complications. Diet counselling and optimum dietary modifications help in the management of obesity. Dietary management includes restriction of calories, carbohydrates, fats, and protein enabling weight loss in a slow and steady phase. This chapter would highlight different diets and methods to reduce obesity. Some diet plans recommend extreme restriction of the carbohydrate intake without any fat restriction, while others are the reverse and place great emphasis on the restriction of fats. What about the pros and cons of similar extreme diets?

The basis of this discussion has to be identified with the increasing incidence of obesity, correlated effects, and research/guidelines available worldwide (Banerjee 2020; Behl and Misra 2017; Chavan 2021; Dalwai et al. 2013; Joshi and Mohan 2018; Khandelwal et al. 1995; Maheshwar and Rao 2011; McLaughlin et al. 2001; Rabast et al. 1979; Radhakrishna et al. 2010; Ray et al. 2016; Saper et al. 2004; Sauvaget et al. 2008; Sharma et al. 2009; Sherwood et al. 2000; Singh and Singh 2015; Sivasankaran

© The Author(s), under exclusive license to Springer Nature Switzerland AG 2022
S. M. Varghese et al., *Trends in Food Chemistry, Nutrition and Technology in Indian Sub-Continent*, Chemistry of Foods,
https://doi.org/10.1007/978-3-031-06304-6_7

2010; Som and Mukhopadhyay 2015; Srinivasan et al. 1995; Vaidyanathan et al. 2019; Viswanathan et al. 2019). Actually, the discussion should be centred on two different representations of weight increase in the human population worldwide, and not only in the most industrialized countries. In detail, more than 1,900 million adult people were estimated to show a body mass index (BMI) between 25 and 30, and more than 1,300 million people were reported to have only an overweight condition (BMI > 25)(Jane et al. 2015). In other words, approximately 32% of the total overweight + obesity-related people were classified 'obese' (BMI > 30) (Jane et al. 2015; Sarwer and Polonsky 2016), and these adults have a remarkable death and/or chronic disease(s)-associated risk when compared with overweigh people.

It has to be remembered that the number of chronic disorders found and correlated with obesity includes the following diseases (Jane et al. 2015; Sarwer and Polonsky 2016):

(a)   Cardiovascular syndromes,
(b)   Muscular skeletal syndromes,
(c)   Type-2 diabetes,
(d)   Sleep apnea,
(e)   Osteoarthritis,
(f)   Several tumours.

Clinically, both pathological conditions have to be considered as the accumulation of excessive fat materials in the human body because of an energetic imbalance between needed (and really consumed) energy and the amount of kilocalories that are really introduced by an active subject. Actually, there is a need for more research in the area of psychosocial treatments when speaking of obesity and associated co-morbidities. In fact, it has been reported that a notable percentage of obese and extremely obese people can show signs of psychiatric illnesses (Kalarchian et al. 2017; Jones-Corneille et al. 2012; Legenbauer et al. 2009; Mitchell et al. 2012; Rosenberger et al. 2006; Sarwer and Polonsky 2016; Schneider and Mun 2005): depression, typical eating disorders, anxiety, substance abuses, sexual and/or physical abuse, consequences of mental health treatments, etc.

As a consequence, and because of economic impacts on public health structures, some sort of weight-management approaches are used worldwide.

7.2   Management of Body Weight. Possible Options

The main goal of weight-management strategies is to reduce the dimensionally measurable extension of the human body in terms of weight with adequate attention to the balance between nutrient profiles. How could this objective be obtained? The general approach is the reduction of energy intake, and this strategy implies two concomitant actions, as briefly displayed in Fig. 7.1 (Jane et al. 2015; Rosenberger et al. 2006):

Weight-loss Management implies Reduction of Energy Intake…

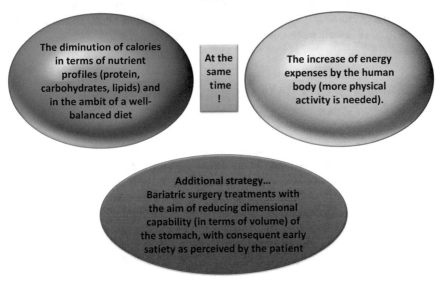

The diminution of calories in terms of nutrient profiles (protein, carbohydrates, lipids) and in the ambit of a well-balanced diet

At the same time !

The increase of energy expenses by the human body (more physical activity is needed).

Additional strategy… Bariatric surgery treatments with the aim of reducing dimensional capability (in terms of volume) of the stomach, with consequent early satiety as perceived by the patient

Fig. 7.1 Weight-loss management implies the reduction of energy intake

(a)   The diminution of calories in terms of nutrient profiles (protein, carbohydrates, lipids) and in the ambit of a well-balanced diet, and

(b)   The increase of energy expenses by the human body (in other words, more physical activity is needed).

Another option (Fig. 7.1) is the possibility of bariatric surgery treatments with the aim of reducing dimensional capability (in terms of volume) of the stomach, with consequent early satiety as perceived by the patient. Weight loss has been reported with good results (Bult et al. 2008; Schneider and Mun 2005). However, this surgical strategy has to be recommended when BMI is ≥ 40 kg/m^2, and also when peculiar conditions different from obesity are present with BMI is ≥ 40 kg/m^2. In addition, important consequences following this type of intervention may be (and should be) expected, such as the increase of bone resorption markers, vitamin D deficiency (with consequent supplementation of calcium and vitamin D), and the occurrence of cholelithiasis (Chapin et al. 1996; Collazo-Clavell et al. 2004; Fujioka 2002; Goode et al. 2004; Marceau et al. 2006; Newbury et al. 2003; Weinsier et al. 1995). Consequently, bariatric surgery is not always recommended.

Overweight and obesity are also a matter of social, environmental, and psychological importance. In addition, biochemical features can differ notably depending on the person. As a result, there is not a simple and easily manageable/understandable weight management program (also named 'weight loss' program, for normal patients), but

different types… and some of these diets may have not adequate scientific reliability (with consequent poor results, and—above all—possible health consequences).
Moreover, the claimed result—in terms of weight diminution—is not easily maintained in many situations. Consequently, the management of obese and overweight
patients is not a simple matter, in India and worldwide.

7.3 Management of Body Weight… and Many Public Safety Issues

In the ambit of overweight/obesity disorders, and with the exclusion of surgical
treatments, physical activity and dietary restrictions are substantially the remaining
options. Physical activity is always recommended, but it is not always compatible
with modern lifestyles worldwide with reference to adults/elderly people and especially to children (Hills et al. 2011; Janssen et al. 2004; Zabinski et al. 2003). With
exclusive reference to recommended diets, the modern situation shows the following
'styles' worldwide, and the situation in the Indian sub-continent is not different, as
shown in Fig. 7.2 (Last and Wilson 2006; Samaha et al. 2003; Stanton 2014; Tsai
et al. 2005):

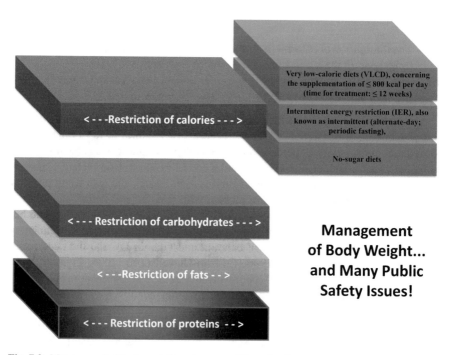

Fig. 7.2 Management of body weight and many public safety issues at the same time worldwide

(a) Calories-restricted diets, such as the 'Atkins diet':

    (a.1) Very-low-calorie diets (VLCD), concerning the supplementation of ≤800 kcal per day (time for treatment: ≤12 weeks);

    (a.2) The intermittent energy restriction (IER), also known as intermittent (alternate-day; periodic fasting), concerning the alternate supplementation of normal calories with short periods of severe restriction (also fasting). Time for treatment: ≤12 weeks. Example: 5 days of regular consumption and 2 days with 500 kcal per day (men) or 600 kcal per day (women);

    (a.3) No-sugar diets such as 'Sugar Busters!'.

(b) Carbohydrates-restricted (low-carb) diets. These treatments limit the amount of supplemented carbohydrates (≤20%, and in the ketogenic diet);

(c) Fats-restricted and protein-restricted diets;

(d) Other 'questionable' diets.

With reference to point a, low-calorie diets are rapid treatments: there is some reported good result when speaking of weight losses, provided that supplementation is nutritionally complete. The problem with rapid treatments is that observed good results appear correlated with water loss (no fat) and reduction of glycogen. Consequently, the real problem—fat excess—remains (Jane et al. 2015; Sarwer and Polonsky 2016). On the other side, VLCD treatments may be useful in some situations provided that close medical supervision is assured (reported successes may be obtained after 12 or more months). However, prevention of weight regain is a concrete problem, suggesting that calories-restricted diets are not the only solution (Sumithran and Proietto 2008). With concern to no-sugar diets such as 'Sugar Busters!', all kinds of foods containing sugars are excluded while low-fat dairy products and vegetable oil-containing products are allowed. Anyway, the complete exclusion of sugar (other examples: 'I Quit Sugar', 'Sweet Poison', etc.) has to be considered as an extreme and questionable choice when speaking of simple weight loss, while the reduction or treatment of diabetes might be more evaluated (Stanton 2014).

With concern to low-carb diets, there is little evidence that weight loss is really important if compared with other diets. On the other side, death risks can be enhanced (augmented supplementation of protein and lipids) because of some tumours, cardiovascular diseases, malnutrition episodes (with low adsorption of dietary fibres, Chap. 3), and ketoacidosis (Dashti et al. 2004, 2006; Mutungi et al. 2008). More research is surely needed in this ambit with concern to the additional possibility that similar diets such as the ketogenic diet may become largely available because of the role of non-carbohydrate nutrients in the diet. Protein and saturated lipids are a major concern, and the use of antioxidants and food additives (Chaps. 1 and 4) against rancidity and protein demolition would be requested in this situation with the aim of contrasting predictable effects on safety and public health. In India, the consumption of high-carbohydrates and high-fat foods has increased the amount of diabetes in the population (Viswanathan et al. 2019), and the use of low-carb diets has to be considered as a strategy against diabetes mellitus. Finally, diets based on the restriction of fats and protein have been found useful in diabetes mellitus treatments in addition to weight loss, but there are many concerns in general.

Fat-restricted diets appear to be the 'right' choice when speaking of effective weight loss. In fact, the rapid and 'permanent' increase of weight in humans and animals depends on the argument of fat materials which should serve as energy storage. Their consumption is extremely slow, and each possible weight-management treatment should aim at the real reduction of fats, especially saturated lipids (because of the difficult metabolization of these long chains). On the one side, the only reduction of fat materials is not useful: in fact, the proportion of slowly digestible protein and sugars would inevitably increase with several health effects as mentioned above (and below). On the other side, the supplementation of vegetable fats/oils can be a key factor when speaking of cardiovascular diseases; however, vegetable oils such as extra-virgin olive oil, remain essentially lipids. Consequently, their intake has to be strictly limited such as in the 'Mediterranean Diet' model (Delgado et al. 2017; Kris-Etherton and Krauss 2020; Issaoui et al. 2020a,b).

High-protein diets such as the Dukan diet (high-protein/low carbohydrate/low fat) have been questioned because of the low dietary intake of certain nutrients and dietary fibres above all. This style and other similar diets such as the South Beach Diet model (no sugars, high protein) are based on initial severe restrictions. However, the only initial effect (rapid weight loss) is not followed adequately: in fact, both mentioned diets are difficultly manageable during long temporal periods.

Finally, the use of diets based on excessive water consumption (the Drinking Man's diet), lemon consumption (the Lemon diet), and other questionable bases should be mentioned because of the very questionable results with concern to rapid weight loss (actually, only water and glycogen are lost, while the fat matter remains unattacked). These diets, similar to strategies based on excessive nutritional deprivation of one of the basical profiles (carbohydrates, lipids, and protein), have another common feature: the insufficient intake of dietary fibres and molecules able to fight/contrast ageing processes in the human body, with the additional absence of natural antioxidants (Chap. 1) able to eliminate or reduce the consequence of abnormal fat consumption (Stanton 2014).

In conclusion, the low intake of carbohydrates limits the consumption and absorption of dietary fibres and natural antioxidants, while fat-restricted diets may obtain long-term results when associated with adequate consumption of antioxidants, fibres, and water. Protein-restricted diets generally cause a disproportion between nutrients, with consequences similar to those for carbohydrate-restricted diets. Finally, 'rapid diets' (also comprehending low-carb approaches) can cause all of these problems, and the rapid promise of weight loss cannot be generally assured (similarly to recommended behaviours…) for long-time periods by consumers/patients.

References

Banerjee S (2020) Implementation of the vegan diet among obese hypothyroid housewives living in metro cities—a review. Int Res J Med Sci 8(1):21–24

Behl S, Misra A (2017) Management of obesity in adult Asian Indians. Indian Heart J 69(4):539–544. https://doi.org/10.1016/j.ihj.2017.04.015

Bult MJ, van Dalen T, Muller AF (2008) Surgical treatment of obesity. Eur J Endocrinol 158(2):135–146. https://doi.org/10.1530/EJE-07-0145

Chapin BL, LeMar HJ Jr, Knodel DH, Carter PL (1996) Secondary hyperparathyroidism following biliopancreatic diversion. Arch Surg 131(10):1048–1052. https://doi.org/10.1001/archsurg.1996.01430220042009

Chavan K (2021) Things that you must know about FAD Diets. Nutrition Meets Food Science. Available https://nutritionmeetsfoodscience.com/2021/08/02/things-that-you-must-know-about-fad-diets/. Accessed 15 Dec 2021

Collazo-Clavell ML, Jimenez A, Hodgson SF, Sarr MG (2004) Osteomalacia after Roux-en-Y gastric bypass. Endocr Pract 10(3):195–198. https://doi.org/10.4158/EP.10.3.195

Dalwai S, Choudhury P, Bavdekar SB, Dalal R, Kapil U, Dubey AP, Ugra D, Agnani M, Sachdev HPS (2013) Consensus Statement of the Indian Academy of Pediatrics on integrated management of severe acute malnutrition. Indian Pediatr 50(4):399–404. https://doi.org/10.1007/s13312-013-0111-3

Dashti HM, Mathew TC, Hussein T, Asfar SK, Behbahani A, Khoursheed MA, Al-Sayer HM, Bo-Abbas YY, Al-Zaid NS (2004) Long-term effects of a ketogenic diet in obese patients. Exp Clin Cardiol 9(3):200–205

Dashti HM, Al-Zaid NS, Mathew TC, Al-Mousawi M, Talib H, Asfar SK, Behbahani AI (2006) Long term effects of ketogenic diet in obese subjects with high cholesterol level. Mol Cell Biochem 286(1–2):1–9. https://doi.org/10.1007/s11010-005-9001-x

Delgado AM, Almeida MDV, Parisi S (2017) Chemistry of the Mediterranean diet. Springer International Publishing, Cham. https://doi.org/10.1007/978-3-319-29370-7

Fujioka K (2002) Management of obesity as a chronic disease: nonpharmacologic, pharmacologic, and surgical options. Obes Res 10(Suppl 2):116S–123S. https://doi.org/10.1038/oby.2002.204

Goode LR, Brolin RE, Chowdhury HA, Shapses SA (2004) Bone and gastric bypass surgery: effects of dietary calcium and vitamin D. Obes Res 12(1):40–47. https://doi.org/10.1038/oby.2004.7

Hills AP, Andersen LB, Byrne NM (2011) Physical activity and obesity in children. Brit J Sports Med 45(11):866–870. https://doi.org/10.1136/bjsports-2011-090199

Issaoui M, Delgado AM, Caruso G, Micali M, Barbera M, Atrous H, Ouslati A, Chammem N (2020a) Phenols, flavors, and the mediterranean diet. J AOAC Int 103(4):915–924. https://doi.org/10.1093/jaocint/qsz018

Issaoui M, Delgado AM, Iommi C, Chammem N (2020b) Polyphenols and the Mediterranean Diet. Springer Nature Switzerland AG, Cham. https://doi.org/10.1007/978-3-030-41134-3

Jane L, Atkinson G, Jaime V, Hamilton S, Waller G, Harrison S (2015) Intermittent fasting interventions for the treatment of overweight and obesity in adults aged 18 years and over: a systematic review protocol. JBI Database Syst Rev Implement Rep 13(10):60–68. https://doi.org/10.11124/jbisrir-2015-2363

Janssen I, Katzmarzyk PT, Boyce WF, King MA, Pickett W (2004) Overweight and obesity in Canadian adolescents and their associations with dietary habits and physical activity patterns. J Adolesc Health 35(5):360–367. https://doi.org/10.1016/j.jadohealth.2003.11.095

Jones-Corneille LR, Wadden TA, Sarwer DB, Faulconbridge LF, Fabricatore AN, Stack RM, Cottrell FA, Pulcini ME, Webb VL, Williams NN (2012) Axis I psychopathology in bariatric surgery candidates with and without binge eating disorder: results of structured clinical interviews. Obes Surg 22(3):389–397. https://doi.org/10.1007/s11695-010-0322-9

Joshi S, Mohan V (2018) Pros & cons of some popular extreme weight-loss diets. Indian J Med Res 148(5):642–647. https://doi.org/10.4103/ijmr.IJMR_1793_18

Kalarchian MA, Marcus MD, Levine MD, Courcoulas AP, Pilkonis PA, Ringham RM, Soulakova JN, Weissfeld LA, Rofey DL (2007) Psychiatric disorders among bariatric surgery candidates: relationship to obesity and functional health status. Am J Psychiatry 164(2):328–334. https://doi.org/10.1176/ajp.2007.164.2.328

Khandelwal SK, Sharan P, Saxena S (1995) Eating disorders: an Indian perspective. Int J Soc Psychiatr 41(2):132–146. https://doi.org/10.1177/002076409504100206

Kris-Etherton PM, Krauss RM (2020) Public health guidelines should recommend reducing saturated fat consumption as much as possible: YES. Am J Clin Nutr 112(1):13–18. https://doi.org/10.1093/ajcn/nqaa110

Last AR, Wilson SA (2006) Low-carbohydrate diets. Am Fam Phys 73(11):1942–1948

Legenbauer T, De Zwaan M, Benecke A, Muhlhans B, Petrak F, Herpertz S (2009) Depression and anxiety: their predictive function for weight loss in obese individuals. Obes Facts 2(4):227–234. https://doi.org/10.1159/000226278

Maheshwar M, Rao DR (2011) A matter of looks: The framing of obesity in popular Indian daily newspapers. J US China Med Sci 8(1):30–34

Marceau P, Biron S, Lebel S, Marceau S, Hould FS, Simard S, Dumont M, Fitzpatrick LA (2006) Does bone change after biliopancreatic diversion? J Gastrointest Surg 6(5):690–698. https://doi.org/10.1016/s1091-255x(01)00086-5

McLaughlin T, Abbasi F, Kim HS, Lamendola C, Schaaf P, Reaven G (2001) Relationship between insulin resistance, weight loss, and coronary heart disease risk in healthy, obese women. Metabol Clin Exp 50(7):795–800. https://doi.org/10.1053/meta.2001.24210

Mitchell JE, Selzer F, Kalarchian MA, Devlin MJ, Strain GW, Elder KA, Marcus MD, Wonderlich S, Christian NJ, Yanovski SZ (2012) Psychopathology before surgery in the Longitudinal Assessment of Bariatric Surgery-3 (LABS-3) psychosocial study. Surg Obes Relat Dis 8(5):533–541. https://doi.org/10.1016/j.soard.2012.07.001

Mutungi G, Ratliff J, Puglisi M, Torres-Gonzalez M, Vaishnav U, Leite JO, Fernandez ML (2008) Dietary cholesterol from eggs increases plasma HDL cholesterol in overweight men consuming a carbohydrate-restricted diet. J Nutr 138(2):272–276. https://doi.org/10.1093/jn/138.2.272

Newbury L, Dolan K, Hatzifotis M, Low N, Fielding G (2003) Calcium and vitamin D depletion and elevated parathyroid hormone following biliopancreatic diversion. Obes Surg 13(6):893–895. https://doi.org/10.1381/096089203322618722

Rabast U, Schönborn J, Kasper H (1979) Dietetic treatment of obesity with low and high-carbohydrate diets: comparative studies and clinical results. Int J Obes 3(3):201–211

Radhakrishna KV, Kulkarni B, Balakrishna N, Rajkumar H, Omkar C, Shatrugna V (2010) Composition of weight gain during nutrition rehabilitation of severely under nourished children in a hospital based study from India. Asia Pac J Clin Nutr 19(1):8–13

Ray M, Ghosh K, Singh S, Mondal KC (2016) Folk to functional: an explorative overview of rice-based fermented foods and beverages in India. J Ethnic Foods 3(1):5–18. https://doi.org/10.1016/j.jef.2016.02.002

Rosenberger PH, Henderson KE, Grilo CM (2006) Correlates of body image dissatisfaction in extremely obese female bariatric surgery candidates. Obes Surg 16(10):1331–1336. https://doi.org/10.1381/096089206778663788

Samaha FF, Iqbal N, Seshadri P, Chicano KL, Daily DA, McGrory J, Williams T, Williams M, Gracely EJ, Stern L (2003) A low-carbohydrate as compared with a low-fat diet in severe obesity. New England J Med 348(21):2074–2081. https://doi.org/10.1056/NEJMoa022637

Saper RB, Eisenberg DM, Phillips RS (2004) Common dietary supplements for weight loss. Am Fam Phys 70(9):1731–1738

Sarwer DB, Polonsky HM (2016) The psychosocial burden of obesity. Endocrinol Metab Clin North Am 45(3):677–688. https://doi.org/10.1016/j.ecl.2016.04.01

Sauvaget C, Ramadas K, Thomas G, Vinoda J, Thara S, Sankaranarayanan R (2008) Body mass index, weight change and mortality risk in a prospective study in India. Int J Epidemiol 37(5):990–1004. https://doi.org/10.1093/ije/dyn059

Schneider BE, Mun EC (2005) Surgical management of morbid obesity. Diab Care 28(2):475–480. https://doi.org/10.2337/diacare.28.2.475

Sharma S, Gulati S, Kalra V, Agarwala A, Kabra M (2009) Seizure control and biochemical profile on the ketogenic diet in young children with refractory epilepsy—Indian experience. Seizure 18(6):446–449. https://doi.org/10.1016/j.seizure.2009.04.001

Sherwood NE, Harnack L, Story M (2000) Weight-loss practices, nutrition beliefs, and weight-loss program preferences of urban American Indian women. J Am Diet Assoc 100(4):442–446. https://doi.org/10.1016/S0002-8223(00)00136-X

Singh A, Singh SN (2015) Dietary fiber content of Indian diets. Asian J Pharm Clin Res 8(3):58–61

Sivasankaran S (2010) The cardio-protective diet. Indian J Med Res 132(5):608–616

Som N, Mukhopadhyay S (2015) Body weight and body shape concerns and related behaviours among Indian urban adolescent girls. Pub Health Nutr 18(6):1075–1083. https://doi.org/10.1017/S1368980014001451

Srinivasan TN, Suresh TR, Jayaram V, Fernandez MP (1995) Eating disorders in India. Indian J Psychiatr 37(1):26–30

Stanton R (2014) Popular diets and over-the-counter dietary aids and their effectiveness in managing obesity. In: Gill T (ed) Managing and preventing obesity: behavioural factors and dietary interventions. Woodhead Publishing, Cambridge, Waltham, and Kidlington, pp 257–274

Sumithran P, Proietto J (2008) Safe year-long use of a very-low-calorie diet for the treatment of severe obesity. Med J Aust 188(6):366–368. https://doi.org/10.5694/j.1326-5377.2008.tb01657.x

Tsai AG, Glick HA, Shera D, Stern L, Samaha FF (2005) Cost-effectiveness of a low-carbohydrate diet and a standard diet in severe obesity. Obes Res 13(10):1834–1840. https://doi.org/10.1038/oby.2005.223

Vaidyanathan S, Kuppili PP, Menon V (2019) Eating disorders: an overview of Indian research. Indian J Psycholog Med 41(4):311–317. https://doi.org/10.4103%2FIJPSYM.IJPSYM_461_18

Viswanathan V, Krishnan D, Kalra S, Chawla R, Tiwaskar M, Saboo B, Baruah M, Chowdhury S, Makkar BM, Jaggi S (2019a) Insights on medical nutrition therapy for type 2 diabetes mellitus: an Indian perspective. Adv Ther 36(3):520–547. https://doi.org/10.1007/s12325-019-0872-8

Weinsier RL, Wilson LJ, Lee J (1995) Medically safe rate of weight loss for the treatment of obesity: a guideline based on risk of gallstone—formation. Am J Med 98(2):115–117. https://doi.org/10.1016/S0002-9343(99)80394-5

Zabinski MF, Saelens BE, Stein RI, Hayden-Wade HA, Wilfley DE (2003) Overweight children's barriers to and support for physical activity. Obes Res 11(2):238–246. https://doi.org/10.1038/oby.2003.37

Chapter 8
Nutrition Education and Diet Counselling

Abbreviation

FAO   Food and Agriculture Organisation of the United Nations

8.1 Nutrition Education and Diet Counselling Today

Nutrition education is an essential component in improving the dietary habits and food choices and nutritional status of a population. The basic aim of this book has been to give a heterogeneous and complete overview of the complexity of nutritional education, mainly from the viewpoint of research chemists and technologists in the first part of the work. However, the final objective—in the Indian subcontinent and worldwide—should be the education of normal consumers with reference to available dietary models and related pros/cons (Chap. 7). Consequently, the book has previously discussed several arguments—antioxidants and nutritional meaning (Chap. 1), phytochemical compounds (Chap. 2), health benefits of some non-nutrients (Chap. 3), food additives (Chap. 4), food safety and technological controls (quality control protocols, Chap. 5), and technologies of preservation in the modern industry (Chap. 6), with the aim of correlating all possible evidences and features of health significance with nutritional education and diet counselling. The choice of many papers and researches mainly carried out by Indian researchers has to be considered here as the attempt to demonstrate that the Indian subcontinent has different challenges at present, when speaking of food-correlated disorders and lack of adequate education, similarly to the rest of the World.

© The Author(s), under exclusive license to Springer Nature Switzerland AG 2022   65
S. M. Varghese et al., *Trends in Food Chemistry, Nutrition and Technology in Indian Sub-Continent*, Chemistry of Foods,
https://doi.org/10.1007/978-3-031-06304-6_8

8.2 Nutrition Education. Basic Definitions, Aims, and Proposed Diet Counseling Approaches

Basically, 'nutrition education' should be considered as a possible collection of experiences that can be taught and consequently learned by all possible food consumers with the aim of ameliorating physical well-being in terms of mobility, vision, hearing, mental ability, Consequently, 'learners' should be able to adopt nutrition-related 'protocols' which can be considered—in the advanced stages of human life above all—as real co-medicinal therapies (Adams et al. 2006; Anonymous 2021). These guidelines should not only promote physical activity, even if this feature is one of the needed pillars in the ambit of human well-being. On the other side, a correct dietary style is needed if the following aims have to be obtained (Adams et al. 2006):

(a) Correct balance between good (affordable) nutrition diets and physical activity;
(b) Adequate awareness of nutrition choices and related consequences, above all with concern for modern dietary models and excessive consumption of foods/beverages rich in one class of nutrients only;
(c) And, last but not least, continuous education support to 'learners' (in other terms, there should not be food consumers without adequate health advice).

Consequently, what about the possible approaches for similar arguments, taking into account that modern life in industrialized countries does not favour active learning concerning healthy and sustainable food nutrition (on the other side, advertising often favours fast-food culture)?

In general, nutrition-oriented education should provide the following tools/instruments with reference to adequate management of body weight, physical activity day-by-day, food safety (basics), and the management of alcoholic beverages (Anonymous 2021):

(1) An education module concerning nutritional profiles of foods and beverages as a four-nutrients complex mixture (protein, fat, carbohydrates, and needed water—Chap. 3).
(2) Adequate knowledge and information concerning needed nutrients for a healthy life (in other terms, prevention of chronic diseases such as cardio-vascular disorders, type-2 diabetes, renal illnesses, etc.).
(3) Useful information concerning biochemical and physiological activities carried continuously by the human body, and related influence of nutrients: digestion, metabolism, excretion, etc. It has to be considered that these processes are notably influenced by the intake of nutrient's and some of the non-nutrient profiles we have already discussed in Chap. 3.
(4) Non-nutrient factors influencing eating/nutritional lifestyles. We are explicitly mentioning all possible features intrinsic to the offered food/beverage of food service which can be able to influence the eating behaviour of normal consumers: (4.1) Enhanced colours, texture, smell... correlation with antioxidants and food additives, Chaps. 1 and 4.

(5) Enhanced palatability and technological features such as chewingness, meltability, etc. ... correlation with food technology and quality control, Chap. 5.
(6) Increased desire for some peculiar food type if compared with other healthier foods Correlation with marketing strategies and associated technological approaches, Chap. 2 (phytochemicals-rich foods = enhanced and 'claimed' health) and Chap. 7 (psychological influence on consumer's awareness).
(7) Sustainability issues (good agricultural practices, selection of short-durability products and concomitant awareness, etc.) ... Correlation with the use and related information on food additives, preservatives, and antioxidants (Chaps. 1, 4, and 5).
(8) Economic issues (in terms of the comparison between similar foods of different price values) ... correlation with quality control, industrial practices, and food preservation above all (Chaps. 5 and 6).

All of these information can be provided by means of different tools: online and offline lectures, web presentations without a physically present teacher (recorded webinars, etc.), cooking broadcasted presentations, academic demonstrations, on-the-ground tasting sessions, etc. Interestingly, and in relation to the portion of elderly people, the online or off-line demonstration by means of institutions such as the 'Universities of the Third Age' appear really popular, and certainly, there is a growing interest. The problem, with these approaches and in other situations, is that the success in terms of ameliorated quality of life and consequent well-being is not always demonstrable (Michalczyk et al. 2020; Pracka et al. 2020). These findings seem to suggest the need for a complete amelioration of training offers, as already pointed out by the Food and Agriculture Organisation of the United Nations (FAO) when speaking of lifelong healthy nutrition programs and the involvement of schools, families, and the whole food-related community. This reflection, based on already established and available dietary guidelines and researches in India and worldwide, can help institutions to create good nutrition-oriented environments, which can serve children, adults, separate food consumer categories (including hospitalized people), and older subjects (Aggarwal et al. 2018; Anonymous 2021; Athavale et al. 2020a, b; Avula et al. 2013; Bailey et al. 2018; Eisenberg and Burgess 2015; Hicks and Murano 2016; Bhandari et al. 2004; Joseph et al. 2015; Kapil 2002; Pavithra et al. 2019; Rathi et al. 2017; Salis et al. 2021; Sethi et al. 2003; Shankar et al. 2017; Sharma et al. 2020; Shekar and Latham 1992; Sivakumar et al. 2016).

In this ambit, diet counselling is extremely useful as a needed co-therapy, in India and abroad (Gulati et al. 2017; Kapur et al. 2008; Rajpoot and Makharia 2013; Sivakumar et al. 2016; Thomas et al. 2009). The problem, especially with claimed patients, is that the following factors—individualized diet counselling and self-management training—are extremely difficult activities both for the healthcare professional and the patient him/herself. In fact, while positive features (reduced number of components in a single family; advanced age; high or medium nutrition awareness and frequent relationships with the dietician) tendentially show new perspectives of well-being, the practical use of diet-counselling advice has two obstacles:

(a) The difficulty to take on personalized diets, person by person, in communities, and also in the same family, and

(b) The real challenge of finding reliable training, which can be carried out taking into account that each patient has a personal history different from other people.

As a result, diet counselling has to be really ameliorated in India and abroad, and the discussion cannot be carried out without the consideration of all points and topics we have briefly discussed in this book. More research and tools for improving nutrition education and diet counselling are surely needed nowadays with the aim of improving counselling approaches. The need for a multidisciplinary strategy (from chemistry to microbiology, from technology to law, economics, and marketing science) is a real problem, but it is—it must be—the key factor for winning the challenge for improved nutrition and health worldwide.

References

Adams KM, Lindell KC, Kohlmeier M, Zeisel SH (2006) Status of nutrition education in medical schools. Am J Clin Nutr 83(4):941S-944S. https://doi.org/10.1093/ajcn/83.4.941S

Aggarwal M, Devries S, Freeman AM, Ostfeld R, Gaggin H, Taub P, Rzeszut AK, Allen K, Conti RC (2018) The deficit of nutrition education of physicians. Am J Med 131(4):339–345. https://doi.org/10.1016/j.amjmed.2017.11.036

Anonymous (2021) Nutrition Education. Washington State Department of Social and Health Services, Washington, DC. https://www.dshs.wa.gov/altsa/program-services/nutrition-education

Athavale P, Hoeft K, Dalal RM, Bondre AP, Mukherjee P, Sokal-Gutierrez K (2020a) A qualitative assessment of barriers and facilitators to implementing recommended infant nutrition practices in Mumbai, India. J Health Pop Nutr 39(1):1–12. https://doi.org/10.1186/s41043-020-00215-w

Athavale P, Khadka N, Roy S, Mukherjee P, Chandra Mohan D, Turton BB, Sokal-Gutierrez K (2020b) Early childhood junk food consumption, severe dental caries, and undernutrition: a mixed-methods study from Mumbai, India. Int J Environ Res Public Health 17(22):8629. https://doi.org/10.3390/ijerph17228629

Avula R, Kadiyala S, Singh K, Menon P (2013) The operational evidence base for delivering direct nutrition interventions in India: a desk review. IFPRI iscussion Paper 01299, International food policy research institute (IFPRI), Washington, DC

Bailey C, Garg V, Kapoor D, Wasser H, Prabhakaran D, Jaacks LM (2018) Food choice drivers in the context of the nutrition transition in Delhi, India. J Nutr Educ Behav 50(7):675–686. https://doi.org/10.1016/j.jneb.2018.03.013

Bhandari N, Mazumder S, Bahl R, Martines J, Black RE, Bhan MK (2004) An educational intervention to promote appropriate complementary feeding practices and physical growth in infants and young children in rural Haryana, India. J Nutr 134(9):2342–2348. https://doi.org/10.1093/jn/134.9.2342

Eisenberg DM, Burgess JD (2015) Nutrition education in an era of global obesity and diabetes: thinking outside the box. Acad Med 90(7):854–860. https://doi.org/10.1097/ACM.0000000000000682

FAO (2021) Food and Nutrition Education. Food and Agriculture Organization of the United Nations, Rome. https://www.fao.org/nutrition/education/en/

Gulati S, Misra A, Pandey RM (2017) Effect of almond supplementation on Glycemia and cardiovascular risk factors in Asian Indians in North India with type 2 diabetes mellitus: a 24–week study. Metabol Syndr Relat Disord 15(2):98–105. https://doi.org/10.1089/met.2016.0066

Hicks KK, Murano PS (2016) Viewpoint regarding the limited nutrition education opportunities for physicians worldwide. Educ Prim Care 27(6):439–442. https://doi.org/10.1080/14739879.2016. 1197048

Joseph N, Nelliyanil M, Sharada Rai RBY, Kotian SM, Ghosh T, Singh M (2015) Fast food consumption pattern and its association with overweight among high school boys in Mangalore city of southern India. J Clin Diagn Res 9, 5:LC13–7. https://doi.org/10.7860/JCDR/2015/13103.5969

Kapil U (2002) Integrated child development services (ICDS) scheme: a program for holistic development of children in India. Indian J Pediatr 69(7):597–601. https://doi.org/10.1007/BF0272 2688

Kapur K, Kapur A, Ramachandran S, Mohan V, Aravind SR, Badgandi M, Srishyla MV (2008) Barriers to changing dietary behavior. J Assoc Phys India 56:29–32. http://repository.ias.ac.in/ 80146/

Michalczyk MM, Zajac-Gawlak I, Zając A, Pelclová J, Roczniok R, Langfort J (2020) Influence of nutritional education on the diet and nutritional behaviors of elderly women at the university of the third age. Int J Environ Res Pub Health 17(3):696. https://doi.org/10.3390/ijerph17030696

Pavithra G, Kumar SG, Roy G (2019) Effectiveness of a community-based intervention on nutrition education of mothers of malnourished children in a rural coastal area of South India. Indian J Pub Health 63(1):4–9. https://doi.org/10.4103/ijph.IJPH_383_17

Pracka M, Dziedziński M, Kowalczewski PŁ (2020) The analysis of nutritional habits of the third age students in poznań. Open Agric 5(1):21–29. https://doi.org/10.1515/opag-2020-0003

Rajpoot P, Makharia GK (2013) Problems and challenges to adaptation of gluten free diet by Indian patients with celiac disease. Nutrients 5(12):4869–4879. https://doi.org/10.3390/nu5124869

Rathi N, Riddell L, Worsley A (2017) Food and nutrition education in private Indian secondary schools. Health Educ 117(2):193–206. https://doi.org/10.1108/HE-04-2016-0017

Salis S, Joseph M, Agarwala A, Sharma R, Kapoor N, Irani AJ (2021) Medical nutrition therapy of pediatric type 1 diabetes mellitus in India: unique aspects and challenges. Pediatr Diabetes 22(1):93–100. https://doi.org/10.1111/pedi.13080

Sethi V, Kashyap S, Seth V (2003) Effect of nutrition education of mothers on infant feeding practices. Indian J Pediatr 70(6):463–466. https://doi.org/10.1007/BF02723133

Shankar B, Agrawal S, Beaudreault AR, Avula L, Martorell R, Osendarp S, Prabhakaran D, Mclean MS (2017) Dietary and nutritional change in India: implications for strategies, policies, and interventions. Ann N Y Acad Sci 1395(1):49–59. https://doi.org/10.1111/nyas.13324

Sharma N, Gupta M, Aggarwal AK, Gorle M (2020) Effectiveness of a culturally appropriate nutrition educational intervention delivered through health services to improve growth and complementary feeding of infants: a quasi-experimental study from Chandigarh. India. Plos One 15(3):e0229755. https://doi.org/10.1371/journal.pone.0229755

Shekar M, Latham MC (1992) Growth monitoring can and does work! an example from the Tamil Nadu integrated nutrition project in rural south India. Indian J Pediatr 59(1):5–15. https://doi.org/ 10.1007/BF02760889

Sivakumar V, Jain J, Tikare S, Palliyal S, Kulangara SK, Patil P (2016) Perception of diet counseling among dental students in India. Saudi J Oral Sci 3(1):36–41. https://doi.org/10.4103/1658-6816. 174335

Thomas D, Joseph J, Francis B, Mohanta GP (2009) Effect of patient counseling on quality of life of hemodialysis patients in India. Pharm Pract 7(3):181–184

Printed in the United States
by Baker & Taylor Publisher Services